STUDY GUIDE to accompany

THE HUMAN BODY
CONCEPTS OF ANATOMY & PHYSIOLOGY

by BRUCE WINGERD

prepared by
ROBERT W. BAUMAN, JR., Ph.D.
Department of Biology, Amarillo College

SAUNDERS COLLEGE PUBLISHING
Harcourt Brace College Publishers

Fort Worth Philadelphia San Diego New York
Orlando Austin San Antonio Toronto
Montreal London Sydney Tokyo

Copyright ©1994 by Harcourt Brace & Company

All rights reserved. No part of this publication may be reproduced or transmitted in any form or by any means, electronic or mechanical, including photocopy, recording, or any information storage and retrieval system, without permission in writing from the publisher.

Requests for permission to make copies of any part of the work should be mailed to: Permissions Department, Harcourt Brace & Company, 8th Floor, Orlando, Florida 32887.

Printed in the United States of America.

Robert W. Bauman, Jr., Ph.D.: Study Guide to accompany THE HUMAN BODY: CONCEPTS OF ANATOMY AND PHYSIOLOGY, by Wingerd

ISBN 0-03-072766-9

456 017 987654321

For Michelle, Jennie, Elizabeth, and Jeremy, and in gratitude to Jesus of Nazareth.

TABLE OF CONTENTS

INTRODUCTION	1
CHAPTER 1: INTRODUCTION TO THE HUMAN BODY	5
CHAPTER 2: THE CHEMICAL BASIS OF THE BODY	21
CHAPTER 3: CELLS: THE BASIS OF LIFE	35
CHAPTER 4: TISSUES	57
CHAPTER 5: ORGANS AND SYSTEMS: OVERVIEW OF THE HUMAN BODY	69
CHAPTER 6: THE INTEGUMENTARY SYSTEM	83
CHAPTER 7: THE SKELETAL SYSTEM	99
CHAPTER 8: THE MUSCULAR SYSTEM	125
CHAPTER 9: ORGANIZATION OF THE NERVOUS SYSTEM	149
CHAPTER 10: THE SPECIAL SENSES AND FUNCTIONAL ASPECTS OF THE NERVOUS SYSTEM	187
CHAPTER 11: THE ENDOCRINE SYSTEM	209
CHAPTER 12: THE BLOOD	225
CHAPTER 13: THE CARDIOVASCULAR SYSTEM	241
CHAPTER 14: THE LYMPHATIC SYSTEM	269
CHAPTER 15: THE RESPIRATORY SYSTEM	289
CHAPTER 16: THE DIGESTIVE SYSTEM	307
CHAPTER 17: NUTRITION AND METABOLISM	333
CHAPTER 18: THE URINARY SYSTEM	345
CHAPTER 19: THE REPRODUCTIVE SYSTEM	359
CHAPTER 20: HUMAN DEVELOPMENT AND INHERITANCE	379
APPENDIX: ANSWERS	393

INTRODUCTION

Welcome! You are beginning an exciting study of the fascinating world of human anatomy and physiology. Rarely will you take a course which is so full of useful information, so packed with details, and so demanding for your concentrated effort and time. Most of you are taking this course as a requirement for your chosen field. Look at anatomy and physiology as one more critical step toward fulfillment of your goals.

HOW TO USE THIS BOOK

A study guide is not a substitute for self discipline and hard work. This book is intended to do one thing, **to guide you in your studying.** You, however, must have the will and commitment to learn. Neither your instructor, your family, your friends, nor this book can make you learn. **You must make the decision to learn.**

Once you have made a quality decision to learn, you may use this book to assist you in your studying. Everyone studies and learns in his or her own way. However, there are some things which successful students have in common.

1. **Have a study place.** Find a place to study where you can be uninterrupted. Do not try to study in the cafeteria or student center unless there are special rooms for studying. You will be distracted by other students. Do not try to study with the television. Research has shown that quietness or soft classical music are best for studying. You may like rock and roll, but it makes it more difficult for your brain to concentrate. You do not have to study in the same place every day, but establish a routine.

2. **Have a set time for study.** You should schedule times to study. Plan to study at least one to two hours for every hour you are in class (including laboratory time). Force yourself to keep to your schedule, and accept no interferences. Do not talk with others. Do not answer the door or the phone. Study!

Try to set your study time as close to class time as possible. At the least, read over your notes or listen to a tape recording of the lecture within an hour or two of class time. Do not try to memorize everything or anything. Merely read over the notes you have taken. Such review will enhance your memorization and comprehension later.

Use this Study Guide to plan your study time. Do not attempt to study an entire chapter during one study session. Complete three, four, or more sections each time.

Each student has a preferred study time. Perhaps your best time is late at night when it is quiet. Other students may prefer a time immediately after they get home from class. Still others may find concentration better in the morning. Whenever is best for you, set a routine and keep with it. Study time at the first of the week is usually more efficient than time scheduled near the end. Plan on studying at the same times during the week, though not necessarily the same time each day. Use the time planner on the next page to schedule your week.

WEEKLY PLANNER

Plan time for eating, commuting, classes, work, recreation, as well as study. Schedule study time in thirty to fifty minute blocks with "breaks" of ten minutes in between. During the breaks get up, walk around, take care of biological functions, walk outside, look at the clouds or stars. Reward yourself with a break after each completed block of study.

Fill out the Planner, as well as the rest of this book, in pencil. It will be easier to correct or modify!

	Sunday	Mon	Tues	Wed	Thurs	Fri	Sat
6:00 am							
7:00							
8:00							
9:00							
10:00							
11:00							
12:00 noon							
1:00 pm							
2:00							
3:00							
4:00							
5:00							
6:00							
7:00							
8:00							
9:00							
10:00							
11:00							
12:00 night							

3. **Stay organized.** Have a set time and place to stud,y and have study materials available. Do not waste time searching for your notes, books, paper, pencils, etc. Have a place to keep all these things.

Make lists of things to accomplish each week and each day. Be realistic. Make each item specific and reasonable. A daily list might be:

1. Read lecture notes.
2. Work through first three pages of study guide and the corresponding text.
3. Make up twenty potential exam questions. (To be answered later.)

When you finish an item, check it off your list. In this way you will be organized, stay focused, and have positive feedback on the amount of work you have accomplished. If you have more list than you have time, pare down your lists, or schedule more time. If you have more time than list, review a previous day's work, go through the list a second time, or start on the next day's list.

4. **Ask questions.** Good students ask question when they do not understand. Poor students may feel that their question is "dumb". It has been said that the only "dumb" question is the one which is not asked. If you feel shy, then ask your questions after class, or schedule a time when you can meet with your professor or teaching assistant. Be sure to write your question down if you are going to ask it at a time outside of class.

5. **Study together.** In addition to your personal study time, schedule time to study with others. A small study group meeting once a week can help you over the rough spots and provide you with positive feedback. Each student will have areas of expertise and weakness. You can help each other, but studying together cannot replace individual preparation.

6. **Anticipate exam questions.** Does the professor emphasize certain topics or ideas. Are some things mentioned in lecture and in lab? What words are in bold print or listed in "Key Terms"? One of the most effective study techniques is to prepare a written exam for yourself. As you study, write out potential exam questions. After you study write out the answers from memory. Before an exam, review the learning objectives, key terms, and questions for review included in each chapter of your textbook.

7. **Follow the directions and suggestions in the study guide.** This study guide will help you through the textbook. If the guide instructs you to read a section of a chapter before filling in the blanks, read it first. If the guide suggests that you fill in the blanks while you read, do it. Some concepts are more difficult than others. The study guide will help you through this material. Fill in all blanks using a pencil. Check your answers after you have finished a block of study time. Filling in blanks is not learning. It will assist you to learn, but you must put forth effort to memorize and comprehend.

There are a variety of study methods suggested in this guide. Some may seem elementary or simplistic to you, but they have been proven to work. If you find one suggestion particularly helpful for your learning style, use it more frequently. If a method of study does not seem effective for you, reeplace it with another more effective method. The purpose of a study guide is to guide you into various patterns of learning. Whatever works, do it. Focus on learning, understanding, and applying your new knowledge. Do not think that you are learning merely by filling in the blanks or memorizing. Be able to use your memory to explain concepts and principles.

I have written this guide to help you master anatomy and physiology. I would appreciate your comments, thoughts, and suggestions.

Dr. Robert Bauman, Jr.
2610 S. Harrison
Amarillo, TX 79109

CHAPTER 1: INTRODUCTION TO THE HUMAN BODY

CONTENT MASTERY

With this chapter you are building a foundation. A building with a firm foundation will stand. The better you prepare and study now, the better you will comprehend and understand more challenging concepts in future chapters.

DIVISIONS OF STUDY. One of your goals is to be able to distinguish between anatomy and physiology. Another is to describe the divisions of anatomy. Don't merely memorize a few definitions here. If you were invited to appear on a television show, you would prepare your understanding of a topic well enough to say more than a short definition. That is the way you should rehearse your understanding of this chapter. Make yourself an expert. If you aren't sure of the meaning of a definition, consult another text or dictionary for a broader description. Become "fluent" in the language of this course!

Match each of the terms with the correct definition or statement. Write the word in the blank. Pay attention to spelling because attention to detail at this point will serve you well in the future! A term may be used more than once, or not at all.

_____ 1. an area of study explaining how the body works

_____ 2. an area of study that describes the location of body parts

_____ 3. examines all structures within a given region of the body

_____ 4. a field of study describing the appearance of body parts

_____ 5. the study of body structures plainly seen without magnification

_____ 6. study of tissues

_____ 7. study of microscopic location and appearance of body parts

anatomy
gross anatomy
histology
microanatomy
systemic anatomy
regional anatomy
physiology

_____ 8. the study of how the body manages energy to remain stable

anatomy
gross anatomy
histology
microanatomy
systemic anatomy
regional anatomy
physiology

_____ 9. an approach that studies anatomy of a given system of the body, such as the digestive

_____ 10. "cutting up"

_____ 11. "the study of nature"

BASIC TERMINOLOGY. Match each of the following directional terms with their definitions and examples. Write the correct answer in the blank being attentive, as always, to your spelling.

_____ 12. toward the head end of the body

anterior/ventral
distal
deep/internal
dorsal
inferior/caudal
lateral
medial
posterior/dorsal
proximal
superficial/external
superior/cranial

_____ 13. away from the head end of the body

_____ 14. toward the front

_____ 15. toward the back

_____ 16. away from the midline

_____ 17. toward the surface of the body

_____ 18. away from the surface of the body

_____ 19. toward the structure's origin

_____ 20. away from a structure's point of attachment to the trunk

_____ 21. The nose is ... to the mouth.

_____ 22. The nose is ... to the eyes.

6

_____ 23. The eyes are on the ... surface
_____ 24. The spinal cord is ... to the "windpipe"
_____ 25. The "breastbone" is ... to the shoulder.
_____ 26. The ears are ... to the nose.
_____ 27. A freckle on the skin is ... to the muscle below.
_____ 28. The lungs lie ... to the rib cage.
_____ 29. The thigh is ... to the ankle.
_____ 30. The thumb is ... to the wrist.

anterior/ventral
distal
deep/internal
dorsal
inferior/caudal
lateral
medial
posterior/dorsal
proximal
superficial/external
superior/cranial

STRUCTURAL LEVELS OF ORGANIZATION. As you read this section of the textbook, note the answers for each of the following on a sheet of paper. When you have finished, go back and try to fill in the blanks without the aid of your text or your answers.

31. Another word for a large molecule is a _____.

32. _____ are small units in the body which are responsible for the structures and the functions of life.

33. Molecules such as proteins, fats, carbohydrates, and nucleic acids combine in an organized manner to form _____.

34. When similar cells combine to accomplish a common function such as movement, the result is called a _____.

35. A(n) _____ is comprised of two or more different types of tissues. For example, muscle tissue, blood, nerve tissue, and epithelium make up the stomach.

36. A _____ is an organization of two or more organs and their associated structures.

THE SYSTEMS OF THE BODY. Using the following names of the systems of the body, identify which system is described in each of the following statements. Write the name of the system in the blanks provided.

cardiovascular	lymphatic	respiratory
digestive	muscular	skeletal
endocrine	nervous	urinary
integumentary	reproductive	

_____ 37. System comprised of pituitary gland, thyroid, adrenal, pancreas, etc.

_____ 38. System which prevents loss of body fluid.

_____ 39. System responsible for support and for protection of softer body parts.

_____ 40. System which includes skin.

_____ 41. System containing bones.

_____ 42. System containing spleen, lymph vessels, and tonsils.

_____ 43. System containing esophagus, liver, and large intestines.

_____ 44. System containing testes, urethra, and penis in the male.

_____ 45. System responsible for removal of dead cells and foreign bodies from body fluids.

_____ 46. System responsible for transportation of materials to and from the cells of the body.

_____ 47. System containing larynx, trachea, and lungs.

_____ 48. System containing muscles that are attached to bones.

_____ 49. System containing brain, spinal cord, and nerves.

cardiovascular lymphatic respiratory
digestive muscular skeletal
endocrine nervous urinary
integumentary reproductive

_____ 50. System primarily responsible for movement of the body.

_____ 51. System influencing homeostasis by stimulating muscles to contract and glands to secrete.

_____ 52. System influencing homeostasis by releasing hormones.

_____ 53. System containing heart, arteries, and veins.

_____ 54. System responsible for transport of materials to and from body cells.

_____ 55. System containing ovaries, uterus, and vagina.

_____ 56. System primarily responsible to maintain homeostasis by controlling water and salt balance in the bloodstream.

_____ 57. System responsible for breaking food into small particles.

_____ 58. System responsible for production of new individuals.

_____ 59. System comprised of kidneys, ureters, urinary bladder, and urethra.

BODY REGIONS. Group the following subdivisions into the major body regions by placing the names of the subdivisions in the appropriate blanks.

SUBDIVISIONS:

abdomen	digits (toes)	pelvis
antebrachium	digits (fingers)	foot (pes)
anterior neck	elbow	posterior neck
axilla	face	shoulder
back	femoral	sole
brachium	gluteal	tarsus
carpus	knee	thorax
cranium	manus	
leg (crus)	palm	

HEAD

60.

61.

NECK

62.

63.

TRUNK

64.

65.

66.

67.

UPPER EXTREMITY

68.

69.

70.

71.

72.

73.

74.

75.

76.

LOWER EXTREMITY

77.

78.

79.

80.

81.

82.

83.

84.

BODY CAVITIES. True-False. Determine the truth of the following statements paying particular attention to the word in italics. If the statement is true, write the word "true" in the blank. If the statement is false, write a word to replace the italicized word that would make the statement true.

_____ 85. The *thoracic* cavity is the single largest cavity in the body.

_____ 86. The ventral cavity is on the *anterior* side of the body.

_____ 87. The thoracic cavity and the abdominopelvic cavity are separated by a thin sheet of muscle known as the *diaphragm*.

_____ 88. The small space between two membranes that surround the heart is called the *pleural* cavity.

_____ 89. The *dorsal* cavity contains a cranial cavity and a ventral canal.

_____ 90. The *mediastinum* contains the heart, thymus, part of the trachea, part of the esophagus, and the major vessels of the heart.

_____ 91. The cavity which contains the urinary bladder, part of the large intestine, and internal reproductive organs is the *abdominal* cavity.

DIAGNOSTIC IMAGING. For each of the following types of diagnostic imaging, write the full name and brief description.

                            FULL TITLE   DESCRIPTION

92. CAT

93. PET

94. Ultrasound imaging (sonography)

95. MRI

CHARACTERISTICS OF LIFE and HOMEOSTASIS.

Match the terms on the left with their descriptions on the right. Each term may have several answers. Place the letter(s) of your choice in the blanks provided. When you are finished, check your work, and match the columns in the reverse order. Could you also fill in a blank with the descriptive word rather than pick it out from a list? Try it, use another sheet of paper, and CHECK YOUR SPELLING!

____ 96. metabolism

____ 97. anabolism

____ 98. catabolism

____ 99. excitability

____ 100. stimuli

____ 101. growth

____ 102. positive feedback

____ 103. negative feedback

____ 104. homeostasis

____ 105. reproduction

a. irritability

b. process by which a single cell divides into two or more

c. literally "the condition of change"

d. [to throw upward]

e. [to throw downward]

f. process by which the body obtains and uses energy

g. process by which the body builds larger molecules (requires energy)

h. a process by which the body breaks apart large molecules to release energy

i. capability of a cell to respond to changes in its environment

j. process by which the environment of the body is kept relatively stable

k. reverses a response to a normal state

l. [similar], [standing still]

m. contraction of the uterus during birth is an example

n. environmental changes

o. an increase in size

12

HEALTH AND DISEASE.

106. Define "disease".

107. What is a lesion?

108. Differentiate between "acute" and "chronic".

109. Define "diagnosis."

110. What does a pathologist do?

111. Distinguish between congenital and immunological diseases.

112. Define "infection."

113. Define "inflammation."

114. Describe "metabolic disease," and give an example.

115. What is neoplastic disease?

INFECTIOUS AGENTS. Complete the following chart to summarize the variety of microscopic organisms that can invade the body and cause disease.

INFECTIOUS AGENT	DESCRIPTION	EXAMPLES
116. Viruses		
117. Bacteria		
118. Fungi		
119. Protozoa		

LABELS AND LISTS

1. Label the following three **basic** types of body planes:

    A. _____

    B. _____

    C. _____

2. On the same illustration, which letter represents a midsagittal plane? ____

3. What is the difference between a midsaggital plane and a parasagittal plane?

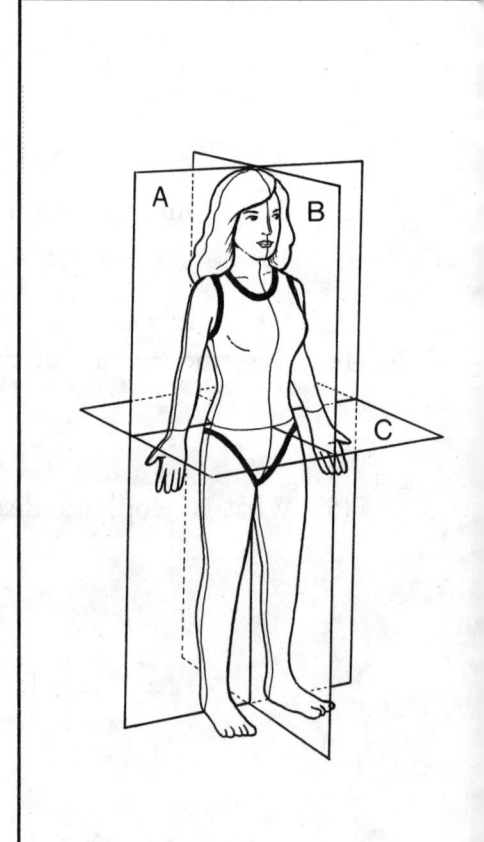

14

4. List the six structural levels of organization of the human body from most simple to the most complex.

   a.   d.

   b.   e.

   c.   f.

5. Name four macromolecules (large molecules) that are structurally fundamental in the body.

   a.   c.

   b.   d.

6. Name the four major types of tissues in the body.

   a.   c.

   b.   d.

7. Name the eleven systems of the human body.

   a.   g.

   b.   h.

   c.   i.

   d.   j.

   e.   k.

   f.

8. Label the following illustration of the major regions of the body, their divisions, and important surface features. Use this list of terms for your answers:

abdomen
ankle
antebrachium (forearm)
axilla (armpit)
back
brachium (upper arm)
carpus (wrist)
cranium

crus (leg)
digits (fingers)
digits (toes)
elbow
face
femoral (thigh)
gluteal (buttocks)
knee

manus
palm
pelvis
pes (foot)
shoulder
sole
tarsus (ankle)
thorax

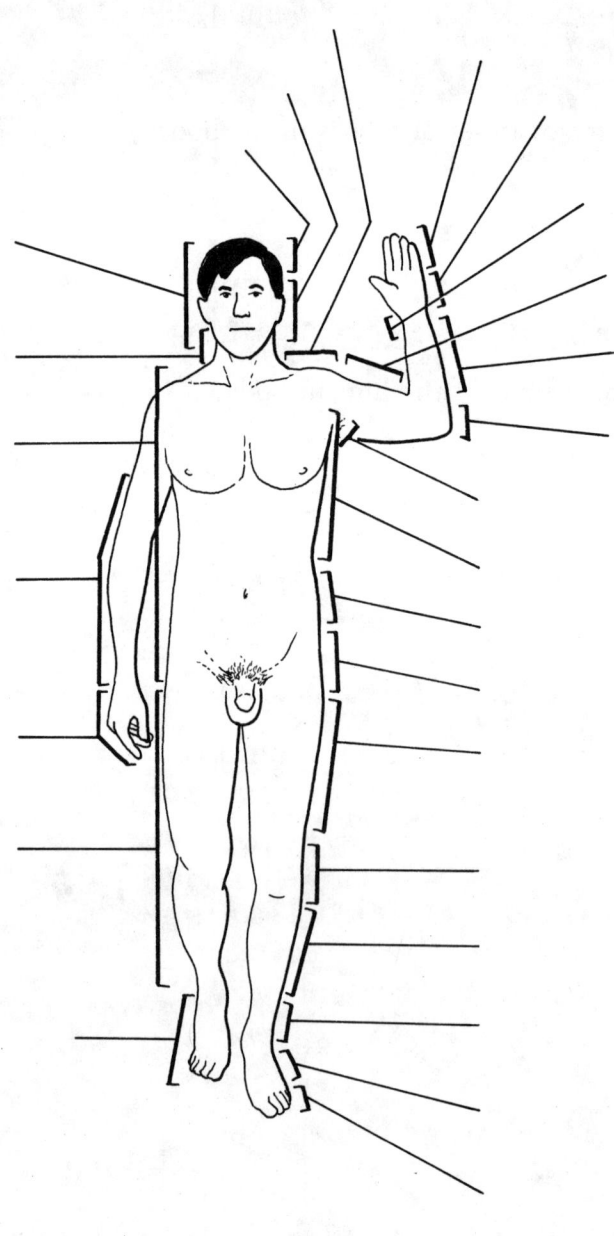

9. Label the following illustration of the body cavities using the list of terms provided.

abdominal cavity
abdominopelvic cavity
abdominopelvic cavity
cranial cavity
diaphragm
dorsal cavity
mediastinum

pelvic cavity
pericardial cavity
pleural cavity
thoracic cavity
ventral cavity
vertebral canal

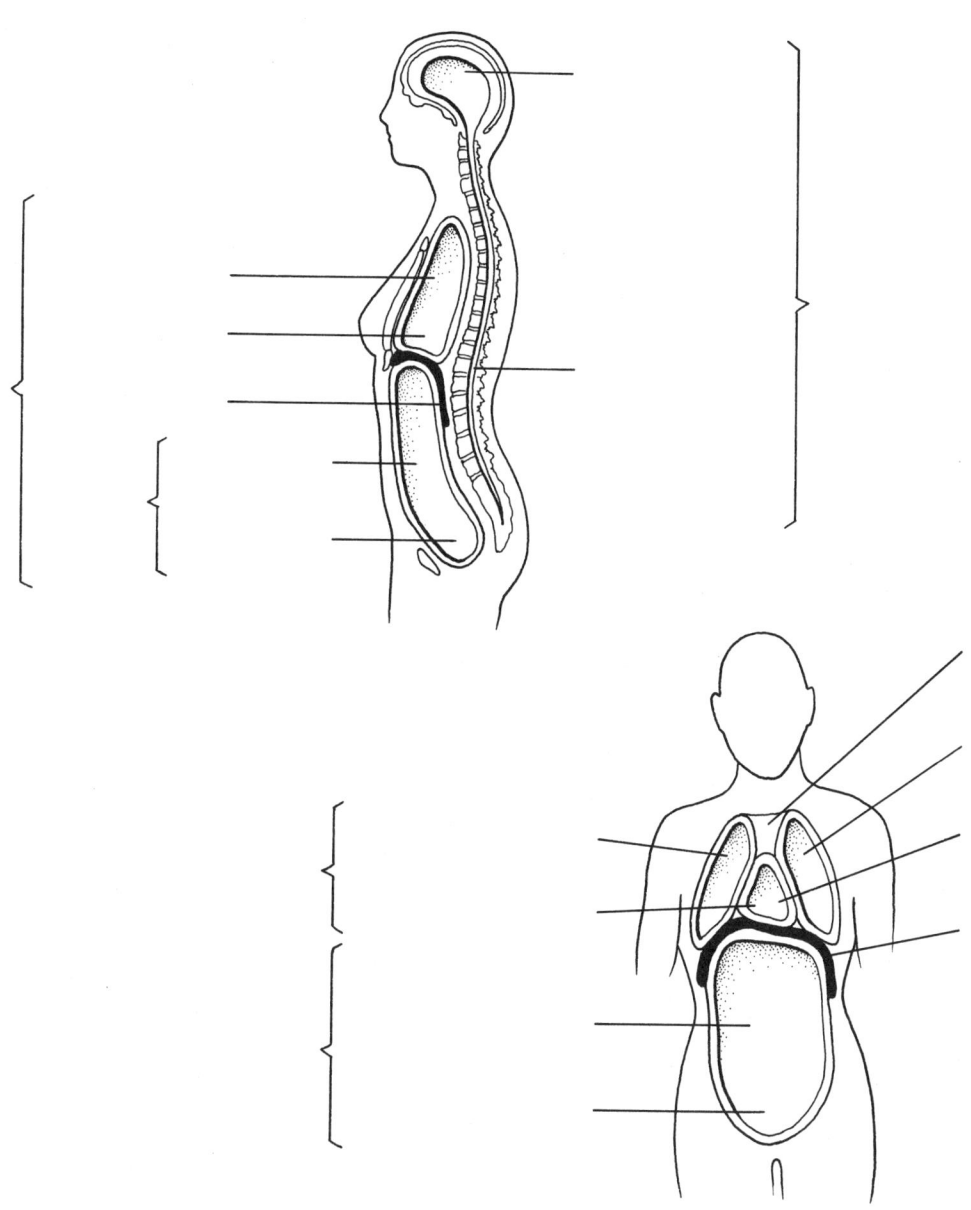

10. Label the following drawings of the divisions of the abdominal region. Note that there are two different ways to subdivide this region. You must be able to label both of them.

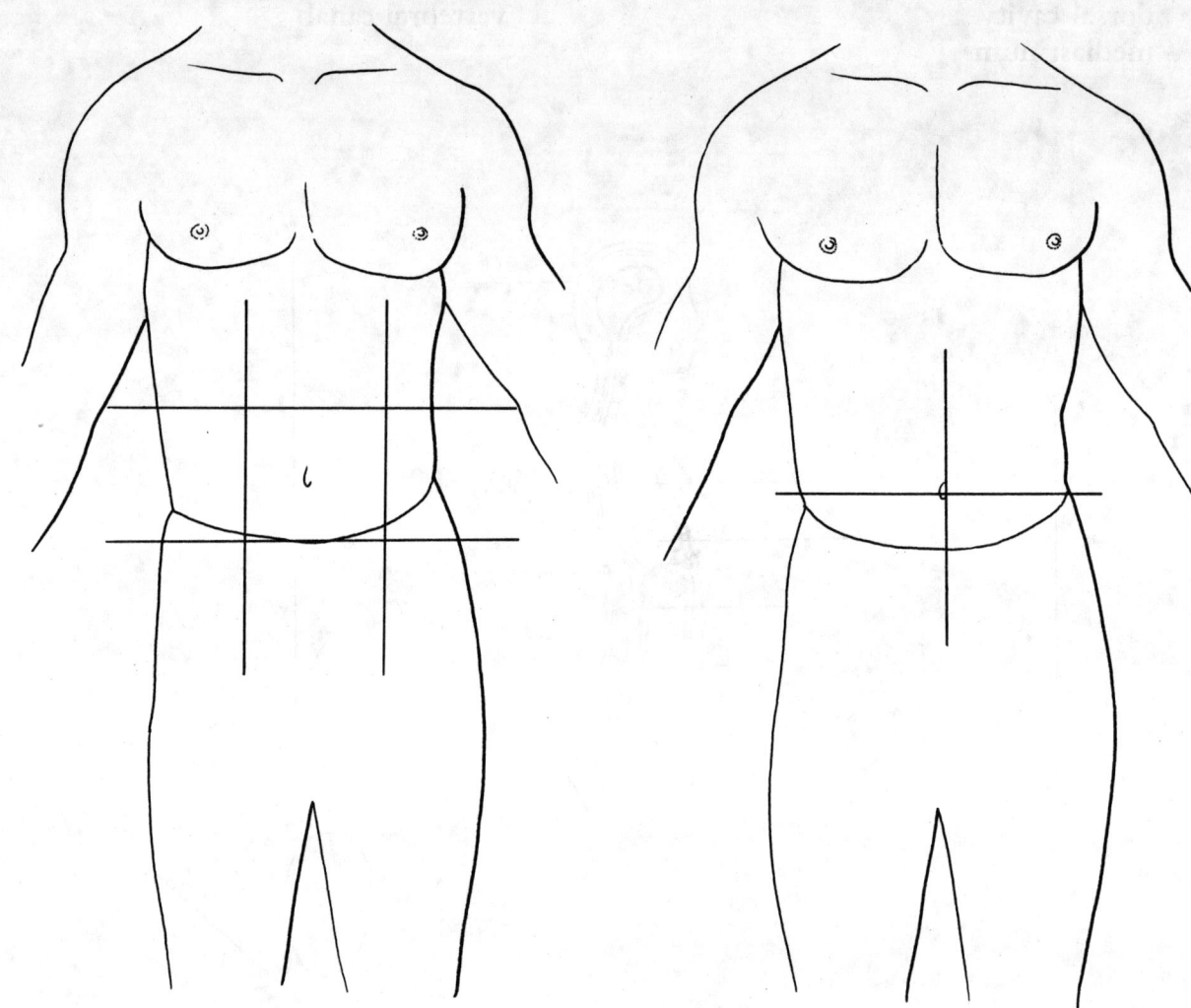

11. List the six characteristics of living cells.

   a.   d.

   b.   e.

   c.   f.

12. List the four types of human disease.

   a.   c.

   b.   d.

ADDITIONAL STUDY

Research indicates that some people learn best when they hear the material they are studying. This may be appropriate for you. Read the Chapter Summary (pages 21 - 22) out loud. Write down the definitions of the KEY TERMS (page 22).

Review the illustrations and know the answers to the questions associated with each one. The answers are on page 23.

Having studied this chapter, close your book, put away your notes, and test yourself by **writing** the answers to the "CONCEPTS CHECKS" and "QUESTIONS FOR REVIEW" in your text (pages 4, 7, 10, 13, 16, 18, 21, 22 - 23). Writing the answers will force you to challenge yourself. If you can write the answers for yourself, you can probably write the answers for your professor.

The day before an exam over this chapter, read the "Learning Objectives", page 2) and review any of the sections which you think will cause you a problem.

CHAPTER ONE REVIEW
by Robert Bauman, Jr., Ph.D.

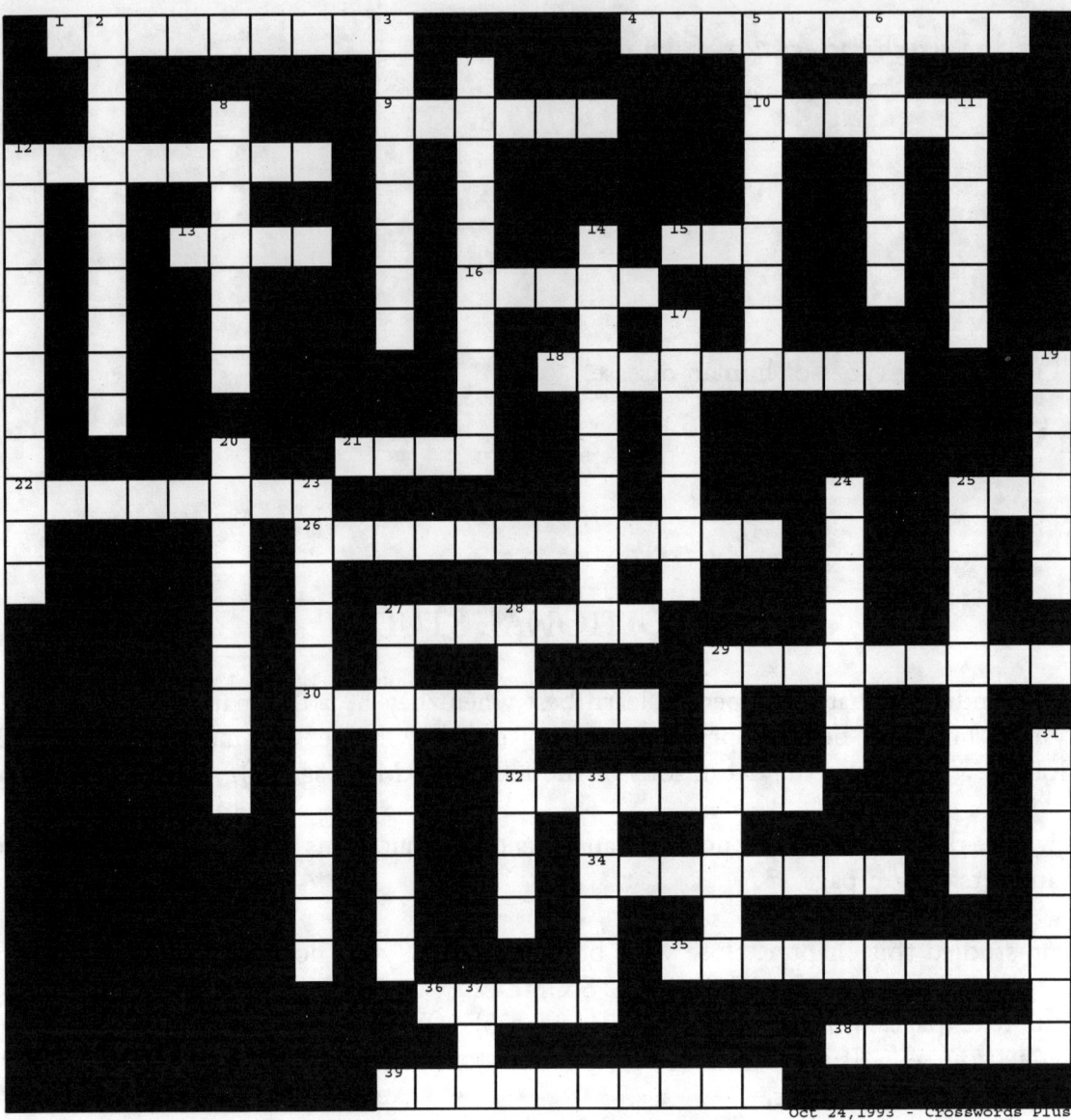

Across

1. invasion by microbes
4. all chemical processes in body
9. increase in size
10. away from origin
12. toward the head
13. internal
15. locating hydrogen with magnetism
16. group of tissues with common purpose
18. study of disease
21. basic unit of life
22. away from the head
25. useful for viewing metabolic activity
26. perception and response to change
27. study of structure
29. toward the back
30. identification of a disease
32. toward the front
34. divides into right and left
35. all chemical reactions in the body
36. short lasting disease
38. toward the midline
39. sonography

Down

2. disease which involoves unchecked cell division
3. common type of feedback
5. cavity containing stomach
6. away from the midline
7. perpendicular section
8. reduction of homeostasis
11. structural change in the body
12. toward the surface
14. large molecules to small molecules
17. long-term disease
19. organs with common purpose
20. study of tissues
23. increase in number
24. divides into anterior and posterior
25. study of function
27. build more comples structures
28. superior ventral cavity
29. change is further promoted
31. toward a structure's origin
33. group of cells with a common purpose
37. computed axial tomography

CHAPTER 2: THE CHEMICAL BASIS OF THE BODY

Some students find this chapter very challenging, because they spend more time moaning about the content than they do mastering the concepts. Rather than approaching this chapter as something strange and formidable, read through the chapter and this study guide with a view to discovering how many things about chemistry you already know. For instance, you have probably heard of carbohydrates, proteins, fats, pH (pH-balanced shampoos), and H_2O (water). See, you already know some chemistry!

Although most of the chemical concepts you will be studying are invisible, their existence has been shown by experiment. Do not let their small size be discouraging. You cannot see atoms or chemical bonds, but there are many things you use every day (such as electricity) which you have never seen. Use your imagination to marvel at the wonders of the chemical world around you!

STUDY HINT. On a separate sheet of paper in your notebook, keep a two-column vocabulary list. In the left column, write each vocabulary word as you encounter it. In the right column, write the definition or give examples. After you have completed the entire section of Content Mastery, come back to your vocabulary list. Compare it to the short definitions given in the Chapter Summary in your textbook by highlighting your accurate definitions or adding to your weak ones. The more you handle these terms, the more comfortable you will be with this very important and foundational aspect of the human body.

CONTENT MASTERY

THE COMPOSITION OF MATTER. Read this section in your text. After each paragraph, answer the pertinent review questions. In this manner, you will work through the reading assignment with concentration.

Complete the following statements with correctly spelled terms from this chapter.

1. Anything that occupies space and has mass is called _____.

2. The amount of matter an object contains is called its _____.

3. The smallest units of matter that have their own distinct sets of properties are _____.

4. A chemical _____ is composed of atoms that share the same characteristics. Oxygen, hydrogen, and carbon are examples.

5. There are _____ (how many?) types of chemical elements which can be studied on a chart called the "Periodic Table of the Elements".

6. The abbreviated name of an element is called a _____ _____.

7. Each atom consists of two basic regions known as a _____ and an _____ shell.

8. The nucleus of an atom contains two basic types of particles: _____ and _____.

9. The particles that surround the nucleus are called _____.

10. Given a proton, a neutron, and an electron, the _____ would be smallest in mass.

11. Given a proton, a neutron, and an electron, the _____ would be the most swiftly moving.

12. Electrons travel within approximate areas of space around a nucleus called _____ _____.

13. Protons have a _____ (+ or -) electric charge.

14. Electrons have a _____ (+ or -) electric charge.

15. The atomic particle with no electrical charge is the _____.

16. In general, the number of positively charged protons equals the number of negatively charged electrons in an atom. This atom is said to be electrically _____.

17. When an atom gains or loses an electron, and thus becomes electrically charged, it is called an _____.

18. The number of protons in each atom is called its _____ _____.

19. The combined total number of protons and neutrons of an atom is almost equal to its _____ _____.

20. A _____ is the result of a combination of two or more atoms.

21. Does the following chemical notation, "H", represent an atom or a molecule? _____.

22. Does the following chemical notation, "O_2" represent an atom or a molecule? _____

CHEMICAL SYMBOLS. Knowledge of chemical symbols will assist you in mastering chemical concepts. Memorize the top thirteen elements and their symbols in Table 2-1. Do not concern yourself with the atomic numbers or percent body mass.

STUDY AID. A helpful mnemonic (memory) device is this "restaurant" sign:

C. HOPKINS CaFe, Mg* with NaCl**

* mighty good!
**Sodium chloride = table salt

The thirteen most common elements in the human body can be found in the sign: **C**arbon, **H**ydrogen, **O**xygen, **P**hosphorus, Potassium (**K**), **I**odine, **N**itrogen, **S**ulfur, **Ca**lcium, Iron (**Fe**), Magnesium (**Mg**), Sodium (**Na**), and Chlorine (**Cl**).

Examine the following chemical symbols carefully and then fill in the blanks with the correct symbol.

_____23. Potassium atom

_____24. Potassium ion

_____25. Calcium ion

_____26. Sodium ion

Ca
Ca^{2+}
K
K^+
Na
Na^+

_____ 27. How many electrons did the sodium ion lose?

_____ 28. How many electrons did the calcium ion lose?

_____ 29. Is a sodium ion a cation or an anion?

_____ 30. The phosphate electrolyte has the chemical symbol PO_4^{3-}. Were electrons gained or lost to make this electrolyte?

_____ 31. How many electrons are involved in the charge on the phosphate electrolyte?

_____ 32. How many oxygen atoms are involved in forming the phosphate electrolyte?

CHEMICAL BONDS. As a preview, read the Chapter Summary concerning this subject. Then read the text for this section, and at the same time complete the following statements by writing correctly spelled terms in the blanks provided.

_____ 33. When molecules are formed by two or more atoms

_____ 34. When outer level electrons are gained, lost, or shared between atoms

_____ 35. An energy relationship between the electrons and protons of the atoms that are joining

_____ 36. A region of space around a nucleus in which electrons exhibit varying amounts of energy

_____ 37. An atom in a stable form

_____ 38. A type of unstable atom

_____ 39. A particle with a negative or positive charge

_____ 40. The joining of a positive ion with a negative ion

_____ 41. Positively charged ions

_____ 42. Negatively charged ions

_____ 43. The sharing of electrons between atoms to fill their outer orbitals

_____ 44. When each atom of a bond contributes a single electron to be shared

_____ 45. When each atom of a bond contributes two electrons to be shared

_____ 46. Weak bonds formed between hydrogen atoms and a molecule with a weak negative charge

CLINICAL USES OF ATOMIC PARTICLES. Fill in the blanks with the most correct answer from the alphabetized list provided here.

cancer   cobalt   isotope   radioisotopes   radioactive decay

47. When an atom contains a different number of neutrons than most other atoms of that element, it is called a(n) _____.

48. Atoms such as the ones discussed in question 47 which are not stable, but rather lose particles from their nuclei at measurable rates are called _____.

49. The loss of subatomic particles is called _____.

50. Radiation can be used to destroy rapidly dividing _____ cells.

51. Compounds labeled with radioisotopes of _____ are often used to monitor absorption of certain vitamins.

52. Radioisotopes (may/may not) _____ be used to detect problems associated with organ functions.

53. X-rays (can/cannot) _____ pass readily through soft tissues.

54. X-rays are used to detect breast tumors. The tumors are (more/less) _____ dense than normal tissue.

CHEMICAL REACTIONS.
55. Distinguish between a synthesis chemical reaction and a decomposition chemical reaction.

56. In the following formula, circle the reactants, and underline the end product.

    glycerol   +   3 fatty acids   ----------------------> fat

57. Why are synthesis reactions important to an adult?

58. The following represents a chemical reaction that occurs during digestion. Circle the end product(s).

    glucose ------------------> carbon dioxide + water

59. Complete the following:

    SYNTHESIS REACTIONS   DECOMPOSITION REACTIONS
    = _____   = CATABOLISM

CHEMICAL COMPOUNDS OF THE CELL. Indicate the validity of each of the following statements by writing "true" in the blanks next to the correct statements. For each false statement, write the word(s) which could best be substituted for the italicized word to make the statement true.

_____ 60. *Inorganic* compounds contain carbon.

_____ 61. *Inorganic* compounds in the body include water, salts, and most acids and bases.

_____ 62. The most important inorganic compound in the body is *carbon*.

_____ 63. The most abundant *inorganic* compound in the body is water.

_____ 64. *Inorganic compounds* are universal solvents.

_____ 65. A liquid that dissolves a solute is called a *solvent*.

_____ 66. The *large* size of water molecules make it ideal for transporting tiny solutes throughout the body.

_____ 67. Water absorbs and releases heat *quickly*.

_____ 68. *Water* is an effective lubricant, reducing friction between moving body parts.

_____ 69. Ions are called "*ionizers*" because of their ability to conduct an electric current.

_____ 70. A *base* is a molecule that releases hydrogen ions when it ionized in water.

_____ 71. A *base* is a molecule that lowers the concentration of hydrogen ions in a solution.

_____ 72. Our bodies *require* acids and bases in order to survive.

_____ 73. The homeostatic mechanisms that keep acid and base concentrations stable are called *pH systems*.

_____ 74. On the pH scale, a value such as "2" would be considered *acidic*.

_____ 75. On the pH scale, a solution that measured "7" could be considered *basic*.

_____ 76. Referring to Table 2-2 in your text, the strongest acid listed is *hydrochloric acid* in stomach.

_____ 77. According to Table 2-2, egg white is *alkaline*.

REVIEW OF INORGANIC CHEMISTRY. Without the use of your textbook, answer the following questions concerning inorganic chemicals.

_____ 78. Which is more acidic, pH 3.5 or pH 2.7?

_____ 79. Which is less basic, pH 11 or pH 12.4?

_____ 80. A scientist discovers a substance which releases hydrogen ions. This substance is a(n) ...

_____ 81. A chemical which reduces hydrogen ion concentration

_____ 82. Name the universal solvent.

_____ 83. In general, which element is **not** found in inorganic compounds?

_____ 84. Salts decompose into ... when placed in water.

_____ 85. A ... system converts strong bases into weak bases and strong acids into weak acids.

_____ 86. What chemical is important as a transport medium?

_____ 87. What chemical is the primary lubricant in the body?

_____ 88. An ion released from a salt in the blood is an ...

_____ 89. Water has a high ___ ___, that is, it absorbs and releases heat slowly.

ORGANIC COMPOUNDS. Read this section of your textbook, then fill the blanks.

_____ 90. General name for compounds which contain carbon

_____ 91. Contains genes

_____ 92. Usually have twice as much hydrogen as oxygen

_____ 93. Triglycerides

_____ 94. Make up 10 - 30% of a cell

_____ 95. Contain only two fatty acid chains

_____ 96. Do not dissolve in H_2O

_____ 97. Contains ribose sugar

_____ 98. Contain three fatty acids joined to glycerol

_____ 99. Building blocks of DNA

_____ 100. Building blocks of RNA

_____ 101. Basic structural material

_____ 102. Structural component of cell membranes

_____ 103. Cholesterol is an example

_____ 104. Long chain of simple sugars

_____ 105. Fats, phospholipids, and steroids

_____ 106. Double helix

_____ 107. Captures and stores energy

_____ 108. Enzymes

_____ 109. Built from nucleotides

ATP
carbohydrates
DNA
fats
inorganic
lipids
nucleic acids
nucleotides
organic
phospholipid
polysaccharide
protein
RNA
steroids

Compare and contrast the two nucleic acids, DNA and RNA, by filling in this table.

	TYPE OF SUGAR	NUMBER OF STRANDS	NUCLEOTIDE BASES PRESENT
DNA	110.	111.	112.
RNA	113.	114.	115.

MORE REVIEW!

_____ 116. The building blocks of proteins are ...

_____ 117. The building blocks of nucleic acids are ...

_____ 118. The building blocks of compounds are ...

_____ 119. The building blocks of atoms are ...

_____ 120. The building blocks of cells are ...

_____ 121. The building blocks of polypeptides are ...

amino acids
atoms
electrons
molecules
neutrons
nucleotides
protons

LABELS AND LISTS

1. Label this planetary model of an atom.

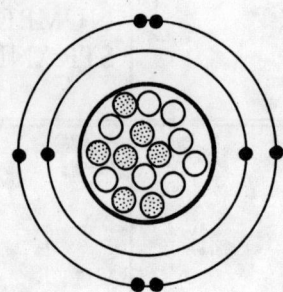

2. Which elements constitute 96% of the body? Write the names and the symbols.

    a.   b.

    c.   d.

3. Which elements constitute 96% of the body? Write the names and the symbols.

    a.   b.

    c.   d.

    e.   f.

4. List three kinds of chemicals bonds.

    a.   b.   c.

5. What four properties of water make it important for life?

    a.

    b.

    c.

    d.

6. What are the four types of organic chemicals?

   a.   b.

   c.   d.

7. What elements are found in carbohydrates? in lipids? in proteins? in nucleic acids?

   carbohydrates:

   lipids:

   proteins:

   nucleic acids:

REVIEW, REVIEW. Now that you have become familiar with many aspects of chemistry, answer the following, more challenging questions in one or two sentences.

1. How is the planetary model of an atom different from the orbital model?

2. Which model is more scientifically accurate?

3. Why is the planetary model used, even though it is not as accurate?

4. Protons have a positive electrical charge and electrons have a negative charge. How is it that atoms which are composed of protons and electrons are neutral in charge?

5. How is an ion different from an atom?

6. Sodium is an explosive, soft metal, and chlorine is a poisonous gas, yet the compound sodium chloride (table salt) is absolutely necessary for life. How can this be explained?

7. Electrons can have different levels of energy. What is the relationship between the energy level and the electron shell?

8. How are covalent and ionic bonds basically different?

9. Why does milk of magnesia help an upset stomach?

10. Is ATP more similar to a DNA nucleotide or an RNA nucleotide? Why?

STUDY AID: The nucleic acids, DNA and RNA, are extremely important molecules. DNA transmits the genetic code from generation to generation. Both DNA and RNA are involved in protein synthesis (see chapter three). Therefore, it is important that you have an understanding of the chemistry of these molecules.

This simple song has helped thousands of students painlessly remember the structure of nucleotides and the way in which they pair.

The DNA Song

(Sung to the tune of "Row, row, row your boat.")

We love DNA, made of nucleotides,
Sugar, phosphate, and a base, bonded down one side.
Adenine and thymine make a lovely pair.
Guanine without cytosine, would be very bare.

Deoxyribonucleic acid!
RNA is ribonucleic acid.
(Repeat from the first.)

c 1975 Michelle Rohrer (Bauman) Used with permission. All rights reserved. Use of the lyrics without the express written permission of the copyright holder is prohibited.

The song will remind you of the basic structure of nucleic acids, the nucleotide building blocks, bonded together to make chains. DNA has two chains. RNA has one chain. The key to inheritance is found in the fact that nucleotides form specific pairs, adenine pairs with thymine (uracil in RNA), and guanine pairs with cytosine. You will examine how DNA and RNA function to direct protein production in chapter three. Learn the song now and you will be humming it to yourself through the next chapter also!

ADDITIONAL STUDY

Read the chapter summary which begins on page s 43 - 44. Write down the definitions of the KEY TERMS (page44)

Review all of the figures and answers the questions associated with each one. The answers are on pages 45.

Having finished studying the chapter, close your book, put away all of your study aids, and write the answers to the "CONCEPTS CHECKS" and "QUESTIONS FOR REVIEW" as if they were an exam (pages 28 - 29, 32, 34, 36, 42, 44 - 45).

CHEMISTRY
by Robert Bauman, Jr., Ph.D.

Oct 27, 1993 - Crosswords Plus

Across

4. in general, chemicals with carbon
6. contains genes
8. ions of salts
10. ionizes in water
11. building block of protien, - acid
13. combination of atoms
16. building block of RNA
18. combination of different atoms
20. occupies space and has mass
21. fast-moving part of an atom
22. breakdown of molecules into simpler ones
23. reduces concentration of H ions
26. charge on a proton
28. sugars and starches
29. negative ion
30. combination to form a larger molecule
31. adenosine triphosphate
33. weakest of the bonds
14. 109 of these
15. something dissolved
17. insoluble in water
19. contains no chain of carbon
24. protein which speeds up a chemical reaction
25. chain of amino acids
26. atomic number = number of
27. dissolver
28. positive ion
29. releases H ions
32. measure of acids and bases

Down

1. charge on an alectron
2. neutral particle
3. bond which involves sharing
5. smallest unit of an element
7. acid formed from nucleotides
9. fat
12. bond between ions

CHAPTER 3: CELLS: THE BASIS OF LIFE

The cell plays a vital role in the way the human body is put together and in the way it works! You will enjoy studying the inner workings of a single cell, how cells produce proteins for the body, how each cell divides, and how cells allow for human reproduction.

STUDY HINTS: Locate the CHAPTER SUMMARY and use it in this way: First, review each section of reading material by carefully reading the corresponding part of the chapter summary. Then, read the section of text material. If you have trouble focusing your attention, read aloud to yourself, or use your fingers to lead your focus on the page as your read. Finally, review each section by completing the following study guide material. Review, review, review. (Most students need this last part to "groove" the information into their memory!)

CONTENT MASTERY

THE ENVIRONMENT OF THE CELL. Complete each of the following statements using the terms provided. Place your answers in the blanks provided.

The fluid-filled area within cells is known as the (1) environment of the cell, whereas the fluid that surrounds the cells is known as the (2) environment. The latter environment is composed mostly of (3) in which may be found dissolved gases, salts, food particles, and cellular products which have been released by a process called (4). That part of the extracellular environment which is between cells is the (5).

1. _____
2. _____
3. _____
4. _____
5. _____

cytoplasm
extracellular
extracellular fluid (ECF)
intercellular environment
interstitial fluid
intracellular
intracellular environment
intracellular fluid

matrix
nucleus
organelles
plasma
plasma membrane
protoplasm
secretion
support

cytoplasm
extracellular
extracellular fluid (ECF)
intercellular environment
interstitial fluid
intracellular
intracellular environment
intracellular fluid

matrix
nucleus
organelles
plasma
plasma membrane
protoplasm
secretion
support

The two types of ECF in the body are (6) and (7). The type of ECF that is located within blood vessels is (8). The type of ECF that is provides a liquid "connection" from cell to cell is (9). The extracellular environment in some areas of the body such as in bone and cartilage contains dense mats of protein called (10). The result of this matting substance is not transportation between cells but rather structural (11).

Within the cell, the fluid substance is called (12). This fluid is bounded by a membrane called the (13) which creates an internal space known as the (14). The total fluid mixture within the cell is called ICF, or (15). Most of the intracellular environment is filled with a gel-like fluid called (16). Structures within the cytoplasm of the cell are called (17). In addition to the plasma membrane and cytoplasm, the protoplasm of the cell also contains a structure that regulates the activities of the cell. This structure is called the (18).

6. _____

7. _____

8. _____

9. _____

10. _____

11. _____

12. _____

13. _____

14. _____

15. _____

16. _____

17. _____

18. _____

Word study. Using these Greek word parts write the word from this section that is described by each of the following clues.

cellular = of the cell
extra = outer
inter = between
intra = within
plasm = form
proto = first
stitial = to set or place [cells]

_____ 19. fluid inside the cell

_____ 20. fluid outside the cell

_____ 21. fluid between neighboring cells

_____ 22. basic substance of a cell

REVIEW. The terms in this section can be confusing so go over them again by filling in these blanks.

cytoplasm **intracellular environment**
intercellular environment **plasma membrane**
interstitial **protoplasm**
intracellular

_____ 23. Material in the body can be either in the extracellular environment or in the ...

_____ 24. The portion of the extracellular environment between cells is the ...

_____ 25. Extracellular fluid can be between cells, i.e. ..., or in the liquid portion of the blood, the plasma.

_____ 26. Fluid within the cells is ... fluid.

_____ 27. The substance of the cell is ...

_____ 28. The portion of protoplasm which separates the extracellular environment from the intracellular environment is the ...

_____ 29. Most of the intracellular environment is filled with a gel-like substance, ...

CELL STRUCTURE AND FUNCTION. This first section of study questions is an overview of cell structures and their functions. First, read the chapter summary of this section. Second, fill in this chart from the summary. **Do not read the chapter until you have completed this overview chart.**

STUDY CHART - CELL STRUCTURE AND FUNCTION

CELL STRUCTURE	PRIMARY FUNCTION	LOCATION
Plasma membrane	30a. _____	b. _____
Cytoplasm	c. _____	d. _____
Endoplasmic reticulum	e. _____	f. _____
Golgi apparatus	g. _____	h. _____
Mitochondria	i. _____	j. _____
Lysosomes	k. _____	l. _____
Peroxisomes	m. _____	n. _____
Cytoskeleton	o. _____	p. _____
Nucleus	q. _____	r. _____
Nuclear membrane	s. _____	t. _____
Nucleoplasm	u. _____	v. _____
Nucleoli	w. _____	x. _____
Chromatin	y. _____	z. _____

Multiple choice over functions of the plasma membrane: Circle the letter of the answer that best completes each statement.

31. The plasma membrane (a) separates the extracellular and interstitial compartments of the cell. (b) regulates the movement of materials between the intra- and extracellular compartments of the body. (c) is made of approximately equal amounts of lipid and protein molecules, plus some carbohydrates. (d) only b and c are correct. (e) All of the above are correct.

32. Which of the following allows some molecules to pass through freely while limiting passage of other molecules? (a) a selectively permeable membrane (b) a freely permeable membrane (c) an impermeable membrane (d) a picky, permeable membrane

33. The plasma membrane is composed of approximately equal amounts of lipid and protein molecules. Which is NOT true of the lipid portion of the plasma membrane? (a) Phospholipids can be found there. (b) Cholesterol molecules create instability in the lipid portion of the plasma membrane. (c) The lipid bilayer is normally in the liquid state. (d) The lipid portion keeps out molecules such as amino acids that do not dissolve in oil.

34. Which term describes a molecule that is attracted to water? (a) hydrophilic (b) hydrophobic (c) covalent thirst

35. Which term describes the type of protein that is attached to the outside and inside of the cell membrane? (a) interstitial protein (b) integral (c) peripheral (d) ion channels (e) glycoprotein

36. A glycoprotein serves as an attachment site for various chemicals which must distinguish between different types of cells. Which of the following attach to glycoproteins? (a) hormones (b) growth factors (c) antibodies (d) all of the above (e) none of the above

37. Which of the following plasma membrane modifications can be described as tiny, slender projections? (a) flagellum (b) microvilli (c) microflagellum (d) cilia (e) microtubules

38. An excretory cell in a kidney may contain as many as 3,000 of these plasma membrane modifications (a) flagella (b) microvilli (c) microflagella (d) cilia (e) microtubules

39. Diffusion is (a) movement of molecules from a higher concentration to a lower concentration. (b) movement of molecules from a lower concentration to a higher concentration. (c) movement up a concentration gradient. (d) rarely observed in our everyday life.

40. Circle each of the following which is longer than microvilli (more than one answer). (a) flagellum (b) cilia (c) microtubules (d) hairs

41. Movement of large molecules such as glucose from an extracellular area of high sugar concentration to the inside of the cell where the concentration of sugar is much lower requires the help of integral proteins which act as escorts. Which term best describes this process? (a) osmosis (b) diffusion (c) facilitated diffusion (d) active transport (e) microdate

42. Which of the following is true of passive processes that transport material across the plasma membrane? Passive processes are (a) powered by kinetic energy. (b) powered by energy from ATP molecules in chemical reactions.

43. Which of the following plasma membrane modifications can be described as a long, single projection that whips about? (a) flagellum (b) microvilli (c) microflagellum (d) cilia (e) microtubules

44. Osmosis is (a) the movement of water molecules only. (b) dependent upon the presence of a selectively permeable membrane. (c) greatly affected (in terms of its speed) by the magnitude of the concentration gradient (d) all of the above (e) both b and c above.

45. During osmosis, water molecules will always move across a membrane toward the solution that has the (a) highest concentration of solutes. (b) highest concentration of solvents. (c) equal concentrations of solute. (d) equal concentrations of solvents.

46. A hypotonic solution has (a) a solute concentration lower than that of the cytoplasm. (b) a solute concentration higher than that of the cytoplasm.

47. Red blood cells burst when they are placed in contact with (a) a hypotonic solution, (b) a hypertonic solution. (c) an isotonic solution. (d) any salt solution.

48. One of the advantages of moving materials by active transport is (a) carrier proteins are not necessary. (b) the process is not dependent on a concentration gradient. (c) the process has no energy cost. (d) receptor sites are not necessary for the process.

49. Which of the following is true if a red blood cell were placed into a beaker filled with a solution that contains approximately the same concentration of water as the red blood cell? (a) The cell would burst. (b) The solution is called an isotonic solution when compared to the inside of the red blood cell. (c) The cell would collapse. (d) The solution is called a hypertonic solution.

50. When small molecules such as urea are pushed because of blood pressure through the plasma membranes of cells in the kidneys, the process is called (a) active transport. (b) osmosis. (c) filtration. (d) cytosis.

51. Plasma membranes of stationary cells of the small intestine can be described by their location. Moving from the open space of the small intestine to the basement membrane, number the surfaces in the order that they would be encountered. ___ lateral border, ___ free surface, ___ basal surface, ___ basement membrane, ___ lumen

Fill in the following blanks to complete the study of the movement of materials across the plasma membrane. Be sure to check your spelling.

52. Active transport in the cell requires the use of energy which is supplied by molecules of _____.

53. Substances moved across membranes by way of active transport move (with/against) ___ _____ the concentration gradient.

54. In general, the movement of very large particles into and out of the cell requires some action in a process known as _____.

55. Assisted movement of large particles into the cell is a type of cytosis called _____.

56. Assisted movement of large amounts of materials out of the cell is a type of cytosis called _____.

57. When a cell extends an arm-like projection toward some particle, surrounds the particle and takes it into the cell, that projection is called a(n) _____.

58. Another term for "cell-eating" is _____.

59. Another term for "cell-drinking" is _____.

60. In general, cells can rid themselves of excess amounts of material by _____.

61. When the material passing out of the cell by exocytosis is a waste product, the process is known as _____.

62. When the material passing out of the cell by exocytosis is an enzyme or other significant product, the process is known as _____.

STOP! Before you go on, study Table 3-1 which graphically summarizes the various types of movements of materials across the plasma membranes. Cover each column in order and practice filling in the "blanks". Review, review, and review some more!

SHORT ANSWER. Answer the following questions to show your grasp of the functions of the plasma membrane in transporting materials into and out of the cell as studied in questions 31-62 above.

63. Contrast, in terms of energy, passive processes and active processes of moving material through plasma membranes.

64. Show your understanding of a "concentration gradient" by giving examples of processes that describe movement with and against the concentration gradient.

65. What is the difference between diffusion and facilitated diffusion?

66. Why is osmosis like diffusion?

67. Buford Jones is repairing his tin roof on a hot summer day. He sweats profusely and then drinks a gallon of distilled water. What might you expect his tissue cells to do?

68. A student places some red blood cells in a hypotonic solution. Which has a greater concentration of **solute**, the cell or the solution?

69. If you looked at these cells in two hours, what you expect to see?

70. Why is it important for medical personnel to give patients isotonic intravenous solution instead of hypertonic or hypotonic solution?

71. What happens to a bacterium which has been phagocytized? (you may need to come back to this after you read about lysosomes.)

72. How are exocytosis and endocytosis alike?

73. How are exocytosis and endocytosis different?

STUDY TIP: At this time go back to questions 1 - 18, cover the paragraphs and write from memory a definition for each of your answer words.

CYTOPLASM and NUCLEUS. Wait! Don't read the text first! The following study could be called a "reading pacer." It deals primarily with terms and definitions concerning the cytoplasm and the five major types of organelles found in the cytoplasm. The answers are generally found in the order that the material is presented in the text, so you can answer **as you read** the material for the first time.

(74) is the protoplasmic material between the plasma membrane and the nucleus. The gel like fluid of the cytoplasm is called "cell fluid" or (75). Within the cell fluid are tiny bodies called (76).

ER, which is an abbreviation for (77), is a network of branching tubules that allow for transportation of new molecules. The two types of ER are: RER (78) and SER (79). The distinguishing difference between the two involves structures called (80). Cells which synthesize volumes of protein molecules would have which type of ER? (81) An example of a cell type that might contain large amounts of smooth ER would be (82).

Cisternae, enzymes, and secretory vesicles are associated with what organelle? (83) Cells that are highly active in secreting enzymes, hormones, and antibodies contain large amounts of
Golgi apparatus.

(84) are called the powerhouses of the cell. The breakdown process associated with mitochondria is known as (85).

The outer membrane of the (86) is selectively permeable whereas the inner membrane is folded into (87). The function of these folds is to increase the (88) of the membrane so that (89) can occur along their surfaces. In the
matrix of the mitochondrion is a concentration of enzymes and (90) which allows the mitochondrion the ability to synthesize some (91). Mitochondrion are in abundance in (92) cells because of the high energy need there.

74. _____
75. _____
76. _____
77. _____
78. _____
79. _____
80. _____
81. _____
82. _____
83. _____
84. _____
85. _____
86. _____
87. _____
88. _____
89. _____
90. _____
91. _____
92. _____

The organelle bounded by a single lipid bilayer membrane and containing hydrolytic enzymes is the (93). These organelles are often called the (94), and because they can digest the cell itself they can also be called (95) or (96). Another function of lysosomes is the release of (97) by exocytosis to break down materials outside the cell. (98) is an example of an inherited disease of lysosomal impairment. Similar to lysosomes, (99) contain enzymes that detoxify various molecules and play a role in fatty acid digestion.

The (100) is found within the cytoplasm. It is/is not (101) an organelle because it has no selectively permeable membrane. Its function is to provide structure for the cell. Types of cytoskeletal elements found in different cells include (102), (103), and (104).

The (105) is the "control center" of the cell because it contains (106), which controls the synthesis of proteins. Its name means (107). It consists of a (108) which envelopes a sap known as (109). The nuclear membrane contains (110) which permit large molecules to pass between nucleoplasm and cytoplasm. Structures in the nucleoplasm include (111) which are not visible during cell division and (112). Nucleoli may be the site of (113). Chromatin, visible during cell division, organizes into (114) which are believed to be made of a single DNA molecule. Segments of the DNA molecules are called genetic units, or (115).

93. _____
94. _____
95. _____
96. _____
97. _____
98. _____
99. _____
100. _____
101. _____
102. _____
103. _____
104. _____
105. _____
106. _____
107. _____
108. _____
109. _____
110. _____
111. _____
112. _____
113. _____
114. _____
115. _____

MATCHING. To review the cytoplasm and nucleus, match the following columns. If you cannot complete this exercise without assistance, see Table 3-2 for a quick summary of the structure and function of cell parts.

_____	116. Envelopes the entire cell	**chromatin**
		cytoplasm
_____	117. Contain enzymes to digest	**cytoskeleton**
		cytosol
_____	118. Site of protein and lipid synthesis	**endoplasmic reticulum**
		golgi apparatus
_____	119. Usually near nucleus with secretory vesicles near plasma membrane	**lysosomes**
		mitochondria
		nuclear membrane
		nucleoli
_____	120. Everything between plasma membrane and nucleus	**nucleoplasm**
		nucleus
		organelles
_____	121. Gel-like fluid around organelles	**peroxisomes**
		plasma membrane
_____	122. Formed bodies	
_____	123. ATP produced here during cellular respiration	
_____	124. Detoxification function	
_____	125. Tiny tubules without membrane	
_____	126. Contains DNA and RNA	
_____	127. Surrounds nucleoplasm and contains pores	
_____	128. Gel-like fluid within nuclear membrane	
_____	129. Site of RNA synthesis; visible during interphase	
_____	130. Made of DNA	

MORE REVIEW! Cover questions 74 - 115 and write the definitions of the answers from memory.

MATCHING. The following cells in the human body each have a large amount of a particular organelle or process which allow them to function effectively. Match the cell types with the structures or process.

_____ 131. neutrophil (white blood cell which rids the body of bacteria)

_____ 132. endocrine gland cell

_____ 133. pancreas (digestive enzymes)

_____ 134. steroid synthesizing cells

_____ 135. kidney tubules

phagocytosis   **smooth endoplasmic reticulum**
Golgi apparatus   **rough endoplasmic reticulum**
secretory vesicles   **mitochondria**

FOR FUN: CELL RAP ... Get a rhythm going, and try this rap to assist your memory of cellular parts:

The plasma membrane around the cell lets different things in or out real well. Engulfing a glob or "drinking" a fluid, with endocytosis it can really do it.

Lysosomes are suicide sacs. Without Golgi bodies, those sacs they'd lack!

Mitochondria make ATP. They're responsible for energy.

Endoplasmic reticulum attaches each thingy to the other one.

The nucleus is like a brain, surrounded by the nuc' membrane.

In the nucleoplasm float nucleoli, full of RNA, do you know why?

Ribosomes are really keen. That's where the cell makes (clap) protein.

Wrap it all up with a rasm dasm. The whole living thing is protoplasm.

c 1977 Michelle Rohrer (Bauman)
Used with permission. All rights reserved. Use of the lyrics without the express written permission of the copyright holder is prohibited.

PROTEIN SYNTHESIS. Because proteins are vital to life, a close look at the process is awesome! Fill in the blanks as you carefully read your textbook. Be sure you have a grasp of each step before you proceed to the next step in the sequence. Remember "The DNA Song" from chapter two? It will assist you now.

_____ 136. What chemical in the nucleus regulates protein synthesis?

_____ 137. Name two processes involved in protein synthesis.

_____ 138. What is the name of the code carried by DNA?

_____ 139. How many nucleotide base pairs are required to code for one amino acid?

_____ 140. What is the name of this group of base pairs?

_____ 141. The triplet codes that code for one polypeptide chain are known as a ...

_____ 142. What molecule carries the genetic code out of the nucleus to the ribosome protein factory?

_____ 143. What process is used by the cell to make mRNA?

_____ 144. How does RNA (mRNA, tRNA, and rRNA) get out of the nucleus?

_____ 145. What is the name of a set of three nucleotides on mRNA that corresponds to a triplet code of DNA?

_____ 146. What process is used by the cell to interpret the codon message into an amino acid sequence?

_____ 147. The set of three nucleotides on tRNA which binds to the codon of mRNA is known as the ...

_____ 148. Where does translation take place?

_____ 149. How many tRNA molecules are associated with the ribosome at any one time?

_____ 150. Which codon of mRNA causes protein synthesis to cease?

_____ 151. How does DNA regulate protein synthesis?

152. Using the terms "triplet code," "base pairs," and "amino acid", describe how DNA carries a message.

153. What is the relationship between a gene and one complete polypeptide chain (or protein)?

154. Compare mRNA, tRNA, and tRNA in terms of their functions and locations.

155. How do the triplet codes of DNA compare to the codon of RNA?

156. What are the differences between transcription and translation in terms of location, reactants, and products?

CELL DIVISION. Read the text material on cell replacement (mitosis and cytokinesis) and sex cell production (via meiosis). After your careful reading, work through the following multiple choice questions. If you are unsure of an answer, clarify your understanding immediately.

_____157. Which of the following activities are the result of cell division?
a. growth   b. gamete production   c. repair   d. both a and b

_____158. During which of the following time phases does chromatin become visible as chromosomes?   a. anaphase   b. metaphase   c. prophase   d. interphase

_____159. The major function of meiosis is to ensure that each of the resultant cells
a. has the same number and kind of chromosomes as the parent cell.
b. has one half the number of chromosomes as the parent cell.   c. has one half the number and one exact copy of one of each pair of homologous chromosomes from the parent cell.   d. all of the above

_____160. The major function of mitosis is   a. to ensure that two genetically identical daughter nuclei are produced.   b. to ensure that two identical daughter cells are produced.   c. to ensure that transcription and translation occur.   d. to ensure that replication is complete.

_____161. Which of the following is NOT true of mitosis?   a. Mitosis results in daughter nuclei which have the same number of chromosomes as the mother nucleus.
b. Mitosis produces gametes by way of two distinct division sequences.   c. Mitosis refers to a nuclear division.   d. One of the events of mitosis is called metaphase.

_____162. During which of the following phases of nuclear division would one expect to see spindle fibers expanded and chromosomes lined up in the center of the cell?
a. Late interphase   b. Prophase   c. Metaphase   d. Anaphase   e. Telophase

_____163. During which of the following phases of nuclear division would one expect to find chromatids separated at the centromeres and pulled toward opposite poles of the cell?   a. Late interphase   b. Prophase   c. Metaphase   d. Anaphase   e. Telophase

_____164. During which of the following phases of nuclear division might one expect to find a cleavage furrow of the cell?   a. Late interphase   b. Prophase   c. Metaphase   d. Anaphase   e. Telophase

_____165. A product of meiosis is a   a. gamete   b. gonad   c. zygote   d. homologous chromosome   e. synapsis

_____166. Synapsis is   a. a matched pairing of homologous chromosomes.   b. a time in meiosis during which the free ends of the chromosomes contact one another.
c. a time in meiosis during which crossing over can occur   d. All of the above are true of synapsis.

_____167. A word which means the same thing as cell division is
   a. mitosis   b. cytokinesis   c. synapsis   d. meiosis   e. metastasis

_____168. A permanent chemical change in a gene is a
   a. tumor   b. mutation   c. transcription   d. metastasis   e. synapsis

_____169. Random mixing of genes during synapsis is called
   a. crossing over   b. mutation   c. metastasis   d. mitosis   e. cytokinesis

_____170. Meiosis is different from mitosis in that it
   a. results in gametes   b. halves the number of chromosomes   c. involves synapsis
   d. involves crossing over   e. all of the above

STUDY HINT ONE: While it may appear that mitosis and cell division (cytokinesis) are the same thing, **they are not**. They do not even have to occur at the same time! Mitosis is necessary for cytokinesis, but not all mitoses result in cell divisions. For example, the cells of the malaria parasite undergo many mitoses before undergoing a single cytokinesis. Thus, for a time the cell is multinucleate. If cytokinesis is going to occur in mitosis, it begins in late anaphase or telophase. Cytokineses do occur during meiosis.

STUDY HINT TWO: Many students find the acrostic "PMAT" helpful in recalling the stages of mitosis in order: **p**rophase, **m**etaphase, **a**naphase, **t**elophase. Interphase is not part of mitosis, but rather the time of growth and normal metabolism between mitoses. PMAT also works for meiosis except, of course, that meiosis is PMAT PMAT!

STUDY HINT THREE: Some words involved with mitosis are similar sounding. Centrioles are organelles which occur in pairs at each end of the spindle. Centro**meres**, on the other hand, are **merely** the constricted region joining two chromatids.

STUDY HINT FOUR: The word "chromosome" has historically been applied to two structures, depending on the stage of mitosis. A prophase or metaphase chromosome consists of two strands joined at the centromere. Each strand of this chromosome is a chromatid. At anaphase, and during telophase, the chromatids separate. Each former chromatid now has a new name; it is called a chromosome! Thus, a particular prophase chromosome is exactly twice as big as its corresponding telophase chromosome.

LABEL AND LIST

1. What three units of protoplasm make up the cell?

   a.   b.   c.

2. Name three modification of the plasma membrane.

   a.   b.   c.

3. Name four ways that material moves across the plasma membrane passively.

   a.   c.

   b.   d.

4. List five kinds of organelles in the cytoplasm.

   a.   c.   e.

   b.   d.

5. List the four phases of mitosis in the order that the would occur.

   a.   c.

   b.   d.

6. Label the solutions as isotonic, hypertonic, or hypotonic.

   a.   b.   c.

7. Label and identify this organelle.
   a.   b.   c.

   d.

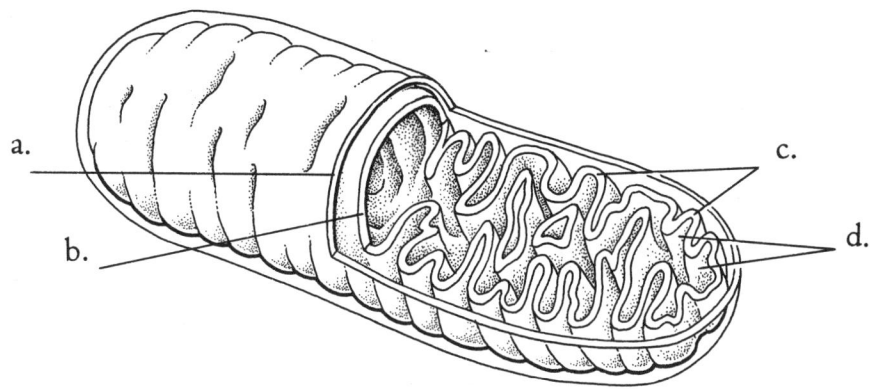

8. Label this generalized diagram of a cell.

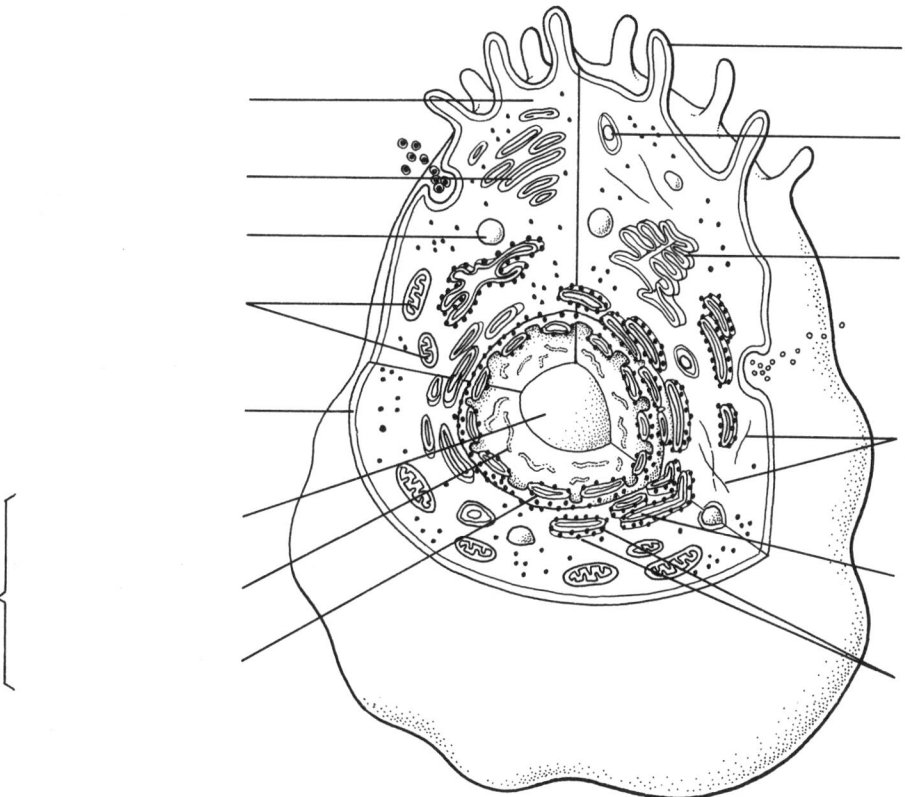

53

9. Label this illustration of a portion of protein synthesis using the following terms: **DNA, mRNA, mRNA processing, nucleolus, nucleus, rRNA, tRNA**

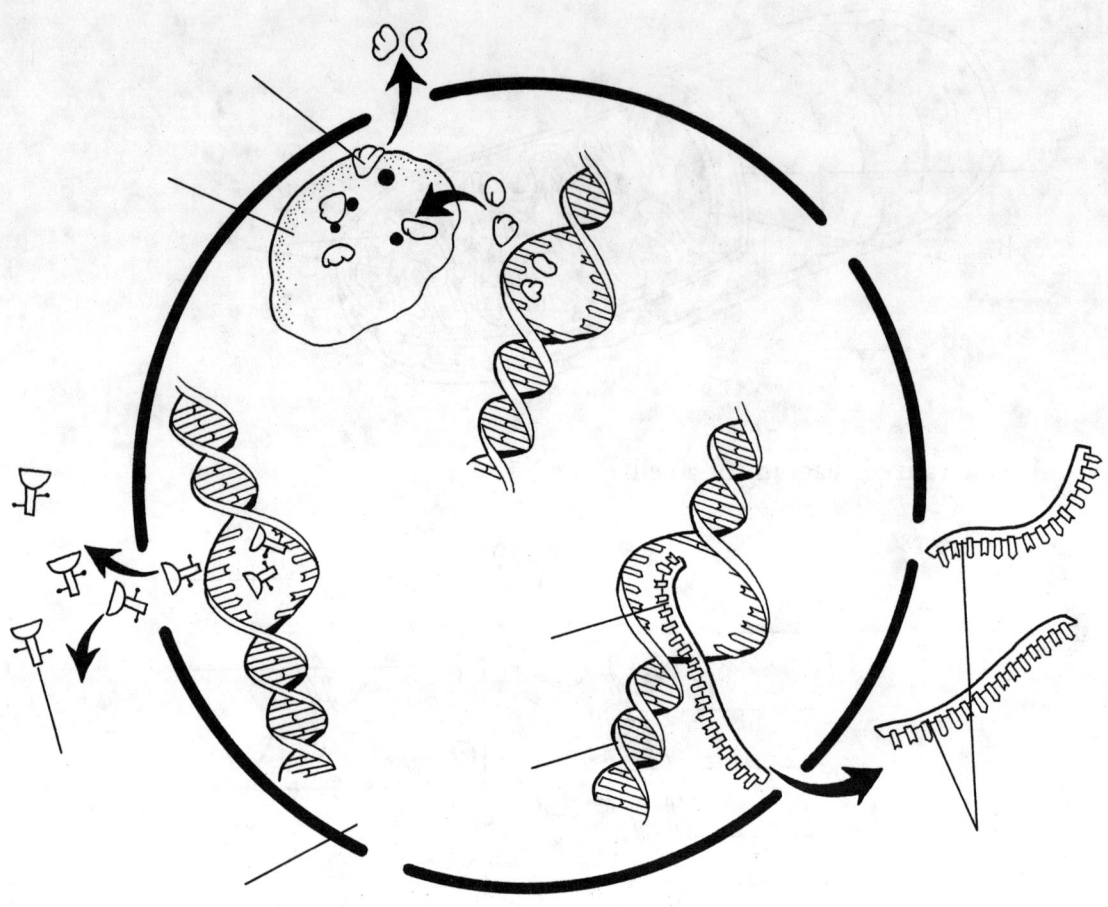

10. Label the stages and structure involved in mitosis and interphase using the following terms: **anaphase, centriole, centromeres, chromatids, chromosome, cytokinesis, early metaphase, early prophase, interphase, late interphase, late metaphase, late prophase, spindle, telophase.**

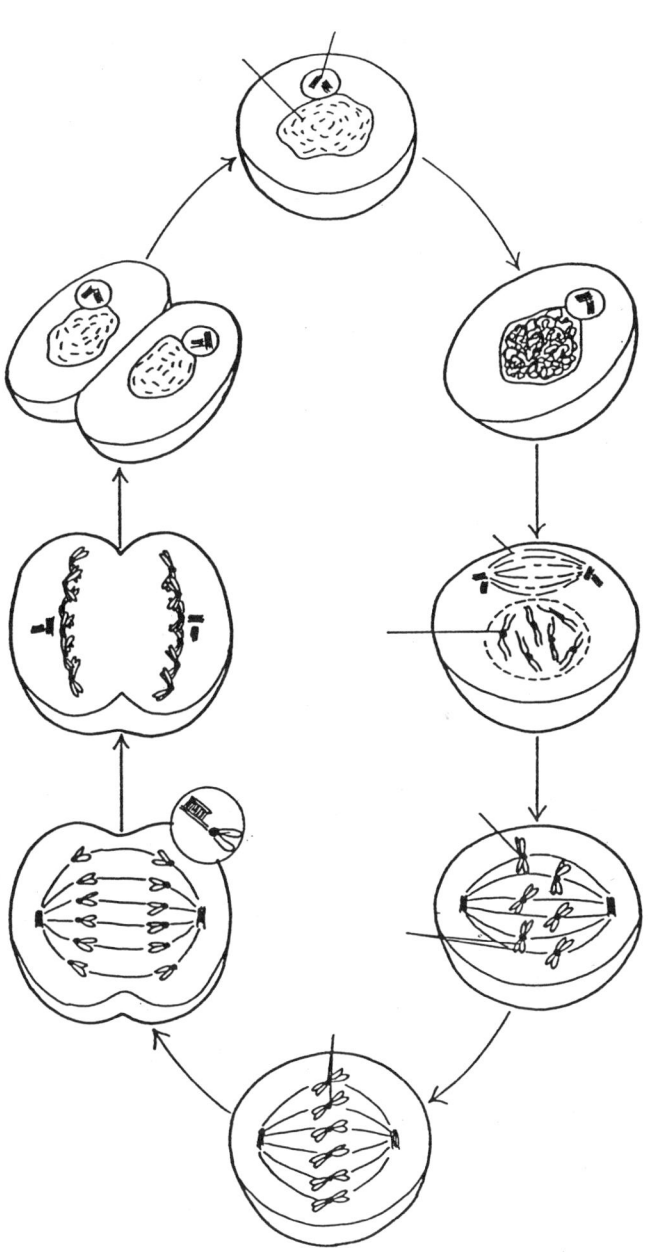

ADDITIONAL STUDY

Read the Chapter Summary (pages 75 - 76) out loud. Write down the definitions of all the KEY TERMS (page 77).

Review the illustrations in your textbook and understand the answers to the questions associated with each one. The answers are on pages 78 - 79 in your textbook.

Having studied this chapter, close your book, put away your notes, and test yourself by **writing** the answers to the "CONCEPTS CHECKS" and "QUESTIONS FOR REVIEW" in your text (pages 48, 49, 53, 54, 59, 63, 65, 70, 74, and 77 - 78). Writing the answers will force you to challenge yourself. If you can write the answers for yourself, you can probably write the answers for your professor.

The day before an exam over this chapter, read the "Learning Objectives", page 46, and review any of the sections which you think will cause you a problem.

CHAPTER 4: TISSUES

There are only four major types of tissue in the human body. These are combined in various ways to form all of the organs and membranes! This chapter explores these tissues (epithelial, connective, muscle, and nerve) which form all the organs you will be studying throughout the course. Your efforts in learning will pay off, so use every learning technique to gain long-term memory of this material.

CONTENT MASTERY

EPITHELIAL TISSUE. When you were a child, you may have enjoyed teasing your friends with the alarm, "EEK! Your epithelium is showing!" Read about this abundant tissue type in your textbook. As you read, fill in the following chart.

TYPE OF COVERING EPITHELIUM	DESCRIBE THE STRUCTURE	EXAMPLES IN THE BODY
Simple squamous		
Simple cuboidal		
Simple columnar		
Stratified squamous		
Pseudostratified columnar		
Transitional		

FILL IN THE BLANKS. Complete the following statements as you study epithelial tissue. Stay alert to your spelling. If a word is very new to you, spell it to yourself several times before going to the next question. Repetition is a key to memorizing effectively.

_____ 1. Because epithelial cells are packed so tightly, blood vessels do not penetrate them; they are said to be ...

_____ 2. The two basic categories of epithelial tissue are covering-lining and ...

_____ 3. Covering and lining epithelial tissue cells are anchored by a layer of intercellular material called ...

_____ 4. The type of epithelium that resembles a tiled floor in top view and is composed of a single layer of flattened cells is ...

_____ 5. The type of epithelium that forms the outermost layer of the skin is ...

_____ 6. The trachea and bronchi are lined with this type of epithelium.

_____ 7. The epithelial tissue known for elasticity and extensibility is ...

_____ 8. ... epithelium is highly specialized to secrete.

_____ 9. ... glands, such as salivary glands, empty their products into ducts.

_____ 10. ... glands, such as the thyroid, have no ducts, but empty their products into the spaces around them where the products then diffuse into the bloodstream.

Review the drawings and photographs of epithelia in your textbook. Be able to identify the various epithelial tissues from a drawing, from a photograph, or from a written description.

CONNECTIVE TISSUE. Before you begin this section of study, read through the types of connective tissue in Table 4-2. The human body is wonderfully held together and supported by a variety of fibers and hardened extracellular material. It would be helpful to reproduce Table 4-2 either by hand or by photocopying it. Add sketches and appropriate notes in the margins to make your study thorough.

Match the terms with their descriptions. Print your answer in the blanks provided. Some terms may be used more than once. Practice correct spelling.

_____ 11. wave-like strands of flexible protein

_____ 12. most common protein in body

_____ 13. main component of scar tissue

_____ 14. scurvy, systemic lupus erythematosus, and rheumatic fever are examples

_____ 15. a disease treated with vitamin C to help the body build collagen

_____ 16. connective tissue fibers composed of elastin

_____ 17. an elastic protein in your skin (not at scar sites)

_____ 18. connective tissue fibers composed of reticulin

_____ 19. the cell in any connective tissue that produces the intercellular material

_____ 20. the most widespread of all connective tissues

_____ 21. also known as areolar

_____ 22. tissue providing insulation padding between organs

adipocytes
adipose tissue
collagenous fibers
collagen
collagen disease
dense connective tissue
dense irregular connective tissue
dense regular connective tissue
elastic fibers
elastin
fibroblast
loose connective tissue
reticular fibers
scurvy

_____ 23. fat cells

_____ 24. tissue contains tightly packed protein fibers and may be regular or irregular

_____ 25. tissue found in dermis containing non parallel arrangement of fibers

_____ 26. tissue containing parallel fibers and found in tendons and ligaments

adipocytes
adipose tissue
collagenous fibers
collagen
collagen disease
dense connective tissue
dense irregular connective tissue
dense regular connective tissue
elastic fibers
elastin
fibroblast
loose connective tissue
reticular fibers
scurvy

Match the following with the correct term from the second list at the right.

_____ 27. protein fibers in a gel substance and may be of three types

_____ 28. dense array of fibers and ground substance

_____ 29. "gristle" cell

_____ 30. small spaces or chambers

_____ 31. dense tissue around cartilage

_____ 32. white-ish cartilage with chondroitin sulfate in matrix

_____ 33. yellowish cartilage with elastic fibers

_____ 34. cartilage noted by thick collagenous fibers in matrix

_____ 35. found in ears, nose, and epiglottis

_____ 36. found at ends of bones at movable joints

_____ 37. found in intervertebral discs

cartilage
matrix
chondrocyte
lacunae
perichondrium
hyaline cartilage
elastic cartilage
fibrocartilage

60

Match the following by placing the correct term in the blanks. The terms may be used more than once.

_____ 38. a type of hard connective tissue with a dense matrix

_____ 39. bone cells

_____ 40. chambers in bone

_____ 41. membrane surrounding a bone

_____ 42. most dense type of bone

_____ 43. haversian canals to allow blood vessels to nourish compact bone

_____ 44. thin, orderly layers of hard matrix in bone

_____ 45. tiny channels through which nutrients reach bone cells

_____ 46. haversian system composed of haversian canal, lamellae, and bone cell inside lacunae

_____ 47. less densely packed type of bone typical of interior of bones

_____ 48. blood-forming tissue in spaces of some bone

_____ 49. thin plates within spongy bone in which are found osteocytes

bone
canaliculi
compact bone
lacunae
lamellae
osteocytes
osteon
osteonic canals
periosteum
red marrow
spicules
spongy bone

BLOOD-FORMING TISSUE AND BLOOD. Mark each statement as "true" or "false". If the statement is false, write the word or words which would replace the italicized words to make it a true statement.

_____50. *Blood-forming tissue* contains stem cells, young blood cells, and protein.

_____51. Blood-forming tissue is the *softest* type of connective tissue in the body.

_____52. *Red marrow* manufactures red blood cells.

_____53. *Hematopoietic* tissue is found in the tonsils of the throat.

_____54. Hematopoietic tissue manufactures all types of blood cells *but two*.

_____55. *Hematopoietic* tissue produces lymphocytes and monocytes.

_____56. Blood is a type of *epithelial* tissue.

_____57. The fluid matrix of blood is known as *plasma*.

_____58. The formed elements of blood include red blood cells, white blood cells, and *plasma*.

_____59. The fibers in the matrix of blood are dissolved *protein* molecules.

Review the figures of connective tissues in your textbook. Be able to identify each type of connective tissue in a photograph, from a written description, or from a drawing.

MUSCLE AND NERVE TISSUE. Multiple Choice. Select the answer which best completes the following statements. Circle the entire answer. After you have finished this section, be sure you can define or describe the circled words. In other words, could you do this section forwards and backwards? Try it until you can.

60. The type of tissue which contains specialized cells which can contract is called
    a. epithelial tissue   b. connective tissue proper   c. blood-forming tissue
    d. muscle tissue   e. nerve tissue

61. The type of muscle tissue attached to bones is called
    a. myosin   b. actin   c. skeletal   d. smooth   e. visceral

62. Which of the following muscle types is under voluntary control?
    a. skeletal   b. smooth   c. cardiac

62. Muscle is attached to bones by way of connective tissue known as
    a. cartilage   b. tendons   c. visceral   d. hyaline

63. Microscopic bands seen on skeletal muscle cells are called
    a. striations   b. filaments   c. tendons   d. viscera

64. The type of muscle tissue found in the walls of the stomach is
    a. skeletal muscle   b. smooth muscle   c. visceral muscle   d. cardiac muscle
    e. both smooth and visceral describe this tissue

65. The type of muscle tissue that lacks striations in its cells is
    a. skeletal   b. visceral   c. cardiac

66. The type of muscle found in the walls of the heart is
    a. skeletal   b. smooth   c. myocardium   d. nervous

67. Cardiac muscle is like skeletal muscle in which of the following features?
    a. voluntary control   b. striations   c. intercalated disks

68. Cardiac muscle is like smooth muscle in which of the following features?
    a. involuntary control   b. striations   c. intercalated disks

69. What tissue has well-developed properties of conductivity and excitability?
    a. skeletal muscle tissue   b. smooth muscle tissue   c. cardiac muscle tissue
    d. nerve tissue

70. A cell in nerve tissue that rapidly conducts electrochemical signals is
    a. neuron   b. neuroglia

71. The most abundant type of cell in the brain is the
    a. neuron   b. neuroglia

TISSUES, TUMORS, AND CANCER. Write your answers to the following questions in complete sentences.

72. What is a tumor?

73. Is there a difference between a neoplastic cell and a tumor cell? If so, what is it?

74. What are carcinogens? Give two examples.

75. What are the differences between benign and malignant tumors?

76. What is metastasis?

77. What is the common name for malignant tumors?

78. What are the early warning signs for cancer?

79. If a concerned friend asked you what she could do to reduce her risk for developing cancer, what could you tell her?

MEMBRANES. Identify which membrane is described by each of the statements. In the blank, write the name of the membrane: **cutaneous, serous, mucous, synovial.**

_____ 80. also known as the skin

_____ 81. associated with joint linings

_____ 82. line internal walls of digestive tract

_____ 83. cover the heart and lung

LABEL AND LIST

1. List the four primary types of tissues in the human body.

    a.   c.

    b.   d.

2. List two types of epithelial tissue.

    a.   b.

3. List four types of connective tissue.

    a.   c.

    b.   d.

4. Complete the chart of the three types of muscle tissue and their characteristics:

Type of Muscle Tissue	Type of Nervous Control	Microscopic Appearance	Example
Skeletal			Muscles of arm
	Involuntary		
		Striated with intercalated disks	

5. Complete the chart of the four types of membranes in the human body and their typical locations.

Membrane	Location
A.	
B.	
C.	
D.	

6. List and describe the three main categories of cancer.

   a.

   b.

   c.

ADDITIONAL STUDY

Read the Chapter Summary (pages 129 - 131). Write out the definitions of all the KEY TERMS (page131).

Review the illustrations in your textbook and understand the answers to the questions associated with each one. The answers are on pages 132.

Having studied this chapter, close your book, put away your notes, and test yourself by **writing** the answers to the "CONCEPTS CHECKS" and "QUESTIONS FOR REVIEW" in your text (pages 114, 118, 120, 123, 126, 128, 131 - 132). Writing the answers will force you to challenge yourself. If you can write the answers for yourself, you can probably write the answers for your professor.

The day before an exam over this chapter, read the "Learning Objectives", page 80, and review any of the sections which you think will cause you a problem.

HISTOLOGY
by Robert Bauman, Jr., Ph.D.

Oct 27,1993 - Crosswords Plus

Across

3. muscle attached to bone
5. cartilage with collagen
7. haversian system
8. conductive tissue
9. dense bone
10. more than one layer
12. cartialge cell
14. return to an original shape
18. bone cell
20. liquid portion of blood
21. elastic protein
24. one layer, looks like more
26. tightly packed connective tissue
27. loose connective
28. channels to feed osteocytes
29. cylindrical
32. contains cells which shorten
33. tissues which suport other body structures
34. lines thoracic and abdominopelvic cavities
35. group of with a common purpose
36. hole within matrix
37. non-epithelial membrane
38. secretes into body fluid

Down

1. one layer
2. lines bladder
3. muscles in stomach
4. secretes into ducts
6. muscle of heart
10. bone with red marrow
11. ring of matrix in compact bone
13. blood forming
15. hair-like extensions
16. closely-packed cells
17. membrane which secretes mucus
19. cube-shaped
22. flat cell
23. haversian canal
25. protein in reticular fiber
27. fat tissue
28. skin
30. tissue with chondrocytes
31. matrix of chondroitin sulfate
32. simple combination of tissues
33. protein fiber in dense tissue

CHAPTER 5: ORGANS AND SYSTEMS: OVERVIEW OF THE HUMAN BODY

Many students feel that they have finally "arrived" when they reach this chapter because it deals with organs and systems that have been familiar for years. The eleven systems of the human body are introduced in this chapter in preparation for detailed study in later chapters. Because you probably have a good platform for learning the material surveyed in this section of study, take this opportunity to "dive in" for fine points of interest. Your efforts will be rewarded because everything you learn will be seen again later in the course!

CONTENT MASTERY

INTEGUMENTARY SYSTEM. Read this section in your textbook before you try to fill in this study guide. Touch the words in bold print with your finger. Say their definitions aloud two or three times. When you have completed this short section of careful reading, try your hand at the following matching questions without consulting the text. Some answers may be used more than once.

blood vessels   glands   hypodermis
cutaneous membrane   homeostasis   receptors
dermis   integument   skin
epidermis

_____ 1. primary organ of the integumentary system

_____ 2. two other ways of naming the "skin"

_____

_____ 3. superficial layer of the skin

_____ 4. deep layer of the skin

_____ 5. a layer of fat and areolar tissue which anchors the skin

_____ 6. forms a waxy protective coat of dead cells and protein over the surface of the body

_____ 7. accessory organs (along with blood vessels and receptors) that are embedded within the dermis

blood vessels　　　　　glands　　　　　　　hypodermis
cutaneous membrane　　homeostasis　　　　receptors
dermis　　　　　　　　integument　　　　　skin
epidermis

_____ 8. may secrete sweat or oil

_____ 9. provide the dermis with enough blood to receive nutrients and carry away waste products

_____ 10. capable of detecting changes in the environment

_____ 11. the overall role of the integumentary system

SKELETAL SYSTEM. Fill in the blanks with the appropriate term concerning the skeletal system. When you check your answers with the key, be sure to check your spelling.

12. The skeletal system consists of _____, joints, and connective tissues.

13. Each bone is regarded as an organ. (true/false) _____

14. The supportive frame of the body is called the _____.

15. The skeleton contains _____ (how many?) bones.

16. The skeleton can be divided into two categories of bones. Those in the vertical axis (down the middle of the body) are part of the _____.

17. Bones lateral to the vertical are considered part of the _____ skeleton.

18. Considering statements "16" and "17" above, the pelvic girdle is found in the _____ skeleton

19. The vertebrae are found in the _____ skeleton.

20. The pectoral girdle is part of the _____ skeleton.

21. The skull is part of the _____ skeleton.

22. Another word for articulation is _____. These terms describe junctions between bones.

23. Of what benefit is the skeletal system to the brain? _____

MUSCULAR SYSTEM. Read the paragraphs that overview the muscular system, and carefully examine the following true-false statements. Write the word "true" or "false" in the blank. Additionally, correct any false statements by changing the word(s) in italics that would make the statement true.

_____24. The organs of the muscular system are the muscles that *attach to the bones*.

_____25. The muscular system is composed of *smooth and skeletal* muscle.

_____26. There are more than *1,000* muscles in the body.

_____27. Skeletal muscle is composed of units known as *contraction groups*.

_____28. Muscles are attached to bone via *tendons*.

_____29. Stimulation for the contraction of muscle cells is provided by the *blood vessels* within the connective tissues.

_____30. The primary function of the muscular system is *homeostasis*.

NERVOUS SYSTEM. Read the pages of text which overview the nervous system. Answer the following multiple choice questions by writing the correct answer in the blank. When you have completed and checked your work, quiz yourself in this way: cover the question, read your answers out loud, and verbally provide the define your answer.

_____ 31. The organs of the nervous systems are structurally classified into the central nervous system and the (afferent nervous system, somatic nervous system, autonomic nervous system, peripheral nervous system)

_____ 32. The nervous system can be classified on the basis of functional differences. It consists of an afferent portion and a portion described as (sensory, efferent, peripheral)

_____ 33. Sensory nerve cells carry information (toward/away from) the central nervous system.

_____ 34. Motor nerve cells carry information (toward/away from) the central nervous system.

_____ 35. The efferent portion of the nervous system that is under conscious control and which stimulates skeletal muscle contraction is called the (somatic nervous system, autonomic nervous system).

_____ 36. The efferent portion of the nervous system that is not under conscious control such as the signaling of the heart and glands is called the (somatic nervous system, autonomic nervous system).

_____ 37. Specialized nerve cells which allow the transmission of impulses throughout the body are generally called (efferent cells, afferent cells, neurons, nervons)

_____ 38. Which of the following is the primary function of the nervous system? (homeostasis, movement, feelings, quick responses, thinking)

ENDOCRINE SYSTEM. After reading the text material, match the following columns describing some of the highlights of the endocrine system. Place your correctly spelled answers in the blanks. This material may be more unfamiliar to you than some of the other systems. Sharpen your attention!

adrenal glands **hormones** **stomach and kidneys**
diffusion **pancreas** **target**
ducts **parathyroid glands** **thymus gland**
gonads **pineal gland** **thyroid gland**
homeostasis **pituitary gland**

_____ 39. small tubes not found in the organs of the endocrine system

_____ 40. an endocrine organ found at base of brain

_____ 41. a large endocrine organ found in the neck

_____ 42. small endocrine organs found in the neck

_____ 43. endocrine organs located on top of each kidney

_____ 44. portions of this organ, located near the stomach, function in the endocrine system

_____ 45. sex glands

_____ 46. an endocrine organ found in the chest which is most active in children

_____ 47. an endocrine organ in the brain

_____ 48. organs which may provide endocrine functions but have primary functions in other systems of the body

_____ 49. the chemicals produced by the endocrine system

_____ 50. process by which hormones move from endocrine glands to the blood stream

_____ 51. the cell affected by a hormone

_____ 52. the overall function of the endocrine system

CARDIOVASCULAR SYSTEM. Since "cardio" refers to the heart and "vascular" refers to blood vessels, guess which organs make up the cardiovascular system. Right! Since so much is generally known about this wonderful system, try this matching section <u>before</u> you read. Even if you score perfectly, read the section in your text.

arteries   formed elements
atria   leukocytes
capillaries   plasma
cardiac muscle   thrombocytes
circulatory system   veins
erythrocytes   ventricles

_____ 53. blood vessels transporting blood away from the heart

_____ 54. blood vessels transporting blood toward the heart.

_____ 55. tiniest blood vessels

_____ 56. one of the two major types of heart tissue

_____ 57. upper chambers of the heart

_____ 58. lower chambers of the heart

_____ 59. chambers of the heart with thinner walls

_____ 60. heart, blood vessels, blood, and lymphatic structures

_____ 61. the collective terms for cells and cell-like particles in the blood

_____ 62. clear, yellowish liquid medium of the blood

_____ 63. red blood cells

_____ 64. white blood cells

_____ 65. platelets

_____ 66. blood cells which carry oxygen and carbon dioxide

_____ 67. blood cells which fight infection

_____ 68. assist in blood clotting

LYMPHATIC SYSTEM. Read the overview material on the lymphatic system before you answer the following questions.

69. In what way is the lymphatic system like the cardiovascular system?

70. In what two ways are the lymphatic and cardiovascular systems different?

    a.

    b.

71. What is the name of the fluid of the lymphatic system?

72. What is produced by lymphoid tissue?

73. What is the primary function of the lymphatic system?

74. In what two ways is the function of the lymphatic system accomplished?

    a.

    b.

75. What is a systemic disease?

76. Give four examples of systemic diseases. (You may use acronyms.)

RESPIRATORY SYSTEM. Read the material which overviews the respiratory system. Label the following statements as true or false. Correct each false statement by replacing the italicized words with the word(s) that will make the statement true.

_____77. One of the primary functions of the respiratory system is to rid the body of dissolved *oxygen* ions.

_____78. Tubes that get air from outside the body to the lungs are called the *conducting zone*.

_____79. Another term for the throat is the *larynx*.

_____80. Another term for the *pharynx* is the voicebox.

_____81. The windpipe is also known as the *trachea*.

_____82. The windpipe branches into two *alveoli*.

_____83. The *respiratory zone* consists of the lungs.

_____84. In the lungs, gases are exchanged between the alveoli and the *capillaries* that surround them.

DIGESTIVE SYSTEM. Surely, the digestive system is one of the first systems to interest students in their younger years of schooling. But remember that the function of this system is the breakdown of food into particles small enough to get to the cells of the other systems. Try filling in the following blanks without looking at your choices or reading the material first. You will be surprised at how knowledgeable you are about the digestive system!

anus	large intestine	salivary glands
bile	liver	small intestine
esophagus	mouth	stomach
gallbladder	pancreas	

_____ 85. where the digestive process begins

_____ 86. produce digestive secretions which empty into the mouth

_____ 87. tube connecting pharynx to stomach

_____ 88. where protein digestion begins

_____ 89. connects stomach to large intestine

_____ 90. a yellowish-green liquid for the digestion of fats

_____ 91. located in the liver area, releases bile

_____ 92. large organ which stores energy-rich molecules

_____ 93. produces digestive enzymes to breakdown protein, carbohydrates, and fats

_____ 94. the organ of the digestive system most responsible for nutrient absorption into the bloodstream

_____ 95. the organ of the digestive system most responsible for water absorption back into the bloodstream

_____ 96. the terminal (last) opening of the digestive system

URINARY SYSTEM. Fill in the blanks as you read the section on the urinary system.

The organs of the urinary system include the two (97) against the posterior wall of the trunk, two (98) that extend downward, a balloon-like (99), and the exiting tube called the (100).

The primary functions of the urinary system are: regulating water, (101), and acid/base balance in body fluids, and removing waste in the form of liquid (102).

97. _____
98. _____
99. _____
100. _____
101. _____
102. _____

REPRODUCTIVE SYSTEM. Match the following terms with their descriptions. Place your correctly-spelled choices in the blanks.

epididymi
fallopian tubes
ovaries
ovum
penis
prostate gland
scrotum
semen

sperm
testes
urethra
uterus
vagina
vas deferens
vulva

_____ 103. the organs that produce male sex cells

_____ 104. male sex cells

_____ 105. cells produced by the testes

_____ 106. external sac covering the testes

_____ 107. carry sperm from epididymis to urethra

_____ 108. carry sperm from ductus deferens to the exterior

_____ 109. reproductive fluid containing sperm cells and liquid

_____ 110. In conjunction with the seminal vesicles and bulbourethral glands, this gland produces the secretions of semen

_____ 111. organs which produce female sex cells

epididymi
fallopian tubes
ovaries
ovum
penis
prostate gland
scrotum
semen

sperm
testes
urethra
uterus
vagina
vas deferens
vulva

_____ 112. female sex cell

_____ 113. site of fertilization

_____ 114. site of embryo implantation and development

_____ 115. female reproductive opening to the exterior

_____ 116. female external genital organs

LABELS AND LISTS

1. List the eleven systems of the human body.

   a.

   b.

   c.

   d.

   e.

   f.

   g.

   h.

   i.

   j.

   k.

2. List five functions of the skeletal system.

   a.

   b.

   c.

   d.

   e.

3. List four functions of the muscular system.

   a.

   b.

   c.

   d.

4. List two structural classifications of the nervous system.

   a.

   b.

5. List five components of the lymphatic system.

   a.

   b.

   c.

   d.

   e.

ADDITIONAL STUDY

Read the Chapter Summary (pages 107 - 109) out loud. Write down the definitions of all the KEY TERMS (page 109).

Review the illustrations in your textbook and understand the answers to the questions associated with each one. The answers are on pages 110 - 111.

Having studied this chapter, close your book, put away your notes, and test yourself by **writing** the answers to the "CONCEPTS CHECKS" and "QUESTIONS FOR REVIEW" in your text (pages 87, 89, 92, 94, 96, 99, 100, and 107). Writing the answers will force you to challenge yourself. If you can write the answers for yourself, you can probably write the answers for your professor.

The day before an exam over this chapter, read the "Learning Objectives", page 112, and review any of the sections which you think will cause you a problem.

82

CHAPTER 6: THE INTEGUMENTARY SYSTEM

OVERVIEW... "Sensitive!" "Vital!" "Productive!" "Colorful!" "Communicative!" "Full of blood, sweat, and tears!" If this sounds like a review of a book worth reading, read on about the integumentary system. Composed of skin, hair, glands, receptors, and blood vessels, the integumentary system protects the body, regulates its temperature, receives stimuli, excretes wastes, and more. It is not merely a hairy sack to live in!

CONTENT MASTERY

FUNCTIONS OF THE INTEGUMENTARY SYSTEM. Read carefully through the description of the five functions of the integumentary system. Check and correct your answers. Polish your memory by writing an answer for each question as if each were an essay or listing question. In other words, don't look at the choices as you write.

_____ 1. Which of the following is NOT a protective function of the skin?
   a. guards against the loss of body fluids
   b. acts as an elastic bag keeping internal organs from being damaged
   c. retards the effects of ultraviolet light
   d. does not permit microorganisms access to the internal areas of the body
   e. All of the above are protective functions of the skin.

_____ 2. In what way does the skin assist in the regulation of body temperature?
   a. Fat layers insulate the body.
   b. Sweat glands allow for evaporative cooling.
   c. Blood vessels can dilate or constrict to regulate the amount of blood which is being cooled or heated by the outside temperature.
   d. All of the above answers are correct.

_____ 3. Which of the following functions of the integumentary system is best described as "excretion?"
   a. the production of a blister to protect sensitive areas
   b. response to ultraviolet light by the production of vitamin D
   c. the release of metabolic wastes through tiny pores
   d. the reaction to heat by the sensory receptors

_____ 4. Which of the following is the major organ of the integumentary system?
   a. hair
   b. glands
   c. sensory receptors
   d. skin

____5. Choose the correctly spelled word from the following list.
   a. entegumentary
   b. integmentery
   c. integumentary
   d. integumentery

____6. What word literally means "upon skin"
   a. epidermis
   b. dermis
   c. hypodermis
   d. subcutaneous

____7. The epidermis is composed primarily of what kind of tissue?
   a. connective tissue
   b. epithelial tissue
   c. muscle tissue
   d. cutaneous tissue

____8. Which of the following is not an accurate description of skin?
   a. Skin is the primary organ of the integumentary system.
   b. Skin consists of two distinct layers.
   c. Skin is also called the cutaneous membrane
   d. Skin is composed of three layers: epidermis, dermis and hypodermis.

____9. The skin is regarded as a membrane as well as an organ.
   a. True, because of the presence of melanocytes.
   b. False, because it is too large and dynamic to be considered a membrane.
   c. True, because of the two layers, epithelium and connective tissue.
   d. False, because it is clearly an organ connecting the outside layers to the inner muscle layers of the body.

____10. The two layers of the skin are
   a. hypodermis and epidermis
   b. endodermis and epidermis
   c. epidermis and dermis
   d. subcutaneous and epidermis

THE EPIDERMIS. Write the word in the blank. A term may be used more than once.

basal cell carcinoma   keratin   epithelium
blood vessels   malignant melanoma   stratum basale
carotene   melanin   stratum corneum
dermis   melanocytes   stratum granulosum
epidermis   pyknosis   stratum lucidum
five   squamous cell carcinoma   stratum spinosum
four   stratified squamous   tanning

_____ 11. literally, bottom layer

_____ 12. literally, spiny layer

_____ 13. literally, transparent layer

_____ 14. literally, horny layer

_____ 15. literally, dark colored cell

_____ 16. waterproof protein of the epithelium

_____ 17. deepest layer of the epidermis

_____ 18. outermost layer of the epidermis

_____ 19. epidermal layer made of prickly looking cuboidal cells with molecular bridges

_____ 20. epidermal layer with twenty to fifty rows of flattened, dead cells

_____ 21. epidermal layer found only in palms and soles of feet

_____ 22. number of layers in epidermis

_____ 23. possible number of weeks from cell production to its sloughing off

_____ 24. cells that secrete a pigment and give skin color

_____ 25. process melanin secretion due to ultraviolet light exposure

basal cell carcinoma	keratin	epithelium
blood vessels	malignant melanoma	stratum basale
carotene	melanin	stratum corneum
dermis	melanocytes	stratum granulosum
epidermis	pyknosis	stratum lucidum
five	squamous cell carcinoma	stratum spinosum
four	stratified squamous	tanning

_____ 26. lethal form of skin cancer due to mutated melanocytes

_____ 27. darkly colored skin pigment

_____ 28. yellowish colored skin pigment

_____ 29. gives Caucasian skin a pink color

_____ 30. a noninvasive, quickly growing cancer

_____ 31. a slowly growing, potentially invasive cancer of the basal layer of the epidermis

_____ 32. a dangerous, quickly metastasizing cancer beginning as a mole

THE DERMIS. Fill in the following blanks as you read the section on the dermis. Upon completion, review your answers before going on.

_____33. The dermis is composed of ___ tissue under the epidermis.

_____34. The intercellular material of the dermis contains a dense mat of ___ which gives the dermis the consistency of a wet sponge.

_____35. Perhaps ___ per cent of the body's blood volume may be present in the dermis.

_____36. The blood supply provides cell nourishment as well as regulating body ___.

_____37. The dermis is divided into two areas (from superficial to deep): the ___ region, and the ___ region.

_____38. The more superficial layer is named for its finger-like projections, or ___.

_____ 39. These projections form contours in the skin surface called ___.

_____ 40. The word "papillary" is derived from the word "papilla" meaning ___.

_____ 41. The word "reticular" is derived from the word "reticula", meaning ___.

_____ 42. The deeper region of the dermis is usually (thicker or thinner?) ___ than the papillary region.

_____ 43. It is composed of dense irregular connective tissue fibers which give the dermis its properties of ___, extensibility, and ___.

_____ 44. A change in the quality of proteins in the reticular region in older adults results in skin ___.

ACCESSORY ORGANS. Answer the following with short but fact-filled answers. Pretend you are the professor filling in an answer key.

45. Why are some accessory organs referred to as "epidermal derivatives" when they are found in the dermis?

46. How is the growth of hair similar to the replacement of the epidermis?

47. Describe the kind of cells found in the shaft of a hair.

48. Describe the action of the arrector pili. Include the advantage of these muscles for fur-bearing animals.

49. What is sebum? What structures secrete it?

50. How is sebum associated with blackheads?

51. Contrast eccrine and apocrine sweat glands.

True-False. Indicate the validity of the following statements by labeling them as "true" or "false." It is an excellent study technique to correct all false statements.

_____52. Hair grows at a normal rate of about 1 inch every 3 days as long as the follicle is healthy.

_____53. Shaving or cutting the hair promotes growth rate.

_____54. About 1,000 hairs are lost and replaced daily in a normal adult scalp.

_____55. Emotional stress can affect baldness.

_____56. Androgens inhibit follicle activity which leads to "male-pattern baldness."

_____57. Hair color is provided by keratin.

_____58. Slower melanin production and increased accumulation of air within the hair shaft promote gray or white hair.

_____59. When "goosebumps" rise on the skin, oil is also forced onto the skin surface.

_____60. Sebaceous glands are not found in the palms and soles.

_____61. Sebum consists of water, fats, cholesterol, protein, and salt.

_____ 62. Sex hormones affect the production of sebum.

_____ 63. Sweat glands are also known as sebaceous glands.

_____ 64. Sweat consists of water, salts, and urea.

_____ 65. Eccrine sweat glands begin functioning during puberty and can cause odoriferous armpits.

_____ 66. Apocrine sweat glands secrete a watery sweat throughout life.

NAILS. After you read the short section which describes nails, match the terms with their description. Pay attention to your spelling. Check your work when you finish. Many people learn best by hearing, so say each word aloud as you review.

brittle nails
eponychium
lunula
nail body
nail root

_____ 67. the visible surface of the nail

_____ 68. another word for cuticle

_____ 69. part of the nail containing the nail matrix

_____ 70. light-colored crescent of the nail

_____ 71. results of calcium deficiency

RECEPTORS. After you read the short section which describes receptors, match the terms with their description. Pay attention to your spelling. Check your work when you finish this section. Test yourself by covering your answers and the matching choices as your review.

away from the brain   Pacinian corpuscles
connective tissue   proximal
distal   sensations
Meissner's corpuscles   toward the brain

_____72. receptors are located on this end of nerve cells

_____73. receptors carry impulses in this specific direction

_____74. nerve impulses are interpreted in the brains as these

_____75. receptors that respond to pressure changes

_____76. receptors that respond to fine touch

SKIN REPAIR. Select the answer which best completes each statement. Write the letter of your choice in the blanks provided.

_____77. Which of the following is least associated with inflammation?
   a. phagocytic white blood cells
   b. epidermal slough off
   c. invading microorganism
   d. fluid accumulation

_____78. What is the functions of a scab?
   a. stimulate stem cells to generate new epidermal tissue
   b. restore the integrity of the epidermis
   c. restrict entry of microorganisms
   d. Both b and c are correct.

_____79. In a minor injury, which occurs first?
   a. repair of the dermis by stem cells
   b. restoration of the epidermis by epithelial cells

_____80. Concerning major injuries, which of the following statements is true?
   a. About one week after the injury, the edges of the wound are pulled together by contraction.
   b. Damaged sweat glands and hair follicles are seldom repaired.
   c. Scar tissue contains repaired hair follicles and muscle cells.
   d. Sebaceous glands are more active for a while.

HYPODERMIS. Indicate your understanding of the hypodermis by labeling the following statements as "true" or writing the word which would correct a false statement.

_____81. The hypodermis lies superior to the skin.

_____82. The hypodermis is composed of loose connective tissue and adipose tissue.

_____83. The hypodermis connects the dermis to the underlying muscle layer by way of collagen.

_____84. One of the functions of the hypodermis is to insulate tissues from temperature extremes, so people living in polar regions may benefit from thick hypodermis.

_____85. One of the benefits of the hypodermis is to provide a shock-absorbing cushion.

HOMEOSTASIS: TEMPERATURE REGULATION. This section of text contains explanations rather than definitions. Most students would rather ask the "how" and "why" questions than explain their answers. Be better than the average student! Perk up your interest as you read, then answer these questions with short but accurate explanations.

86. The narrow range for healthy body temperature is directly related to the nature of enzyme activity. Explain what happens to enzymes when the body temperature moves above or below that healthy range.

87. Suppose you are shingling a roof on a July afternoon in Texas. The temperature outside your body is well above your body's healthy range limit. Sequence the following to show how your body maintains homeostasis:
   a. sweat evaporates and reduces the temperature in skin and nearby blood
   b. brain stimulates sweat glands to secrete
   c. receptors in skin sense the external heat and send message to brain
   d. brain signals blood vessels in dermis to increase blood flow to the skin
   e. cooled blood circulates through body to lower internal temperature
   f. excessive heat in blood passes through the skin to the air

88. Explain why Gertrude Willowfrost's face gets red when she walks to her afternoon summer school class.

89. Explain the cause of shivering.

90. Of what homeostatic value is it to the body to shiver and experience "goosebumps"?

91. Of what homeostatic value is it for the heart to beat faster during an increase in body temperature?

92. When Bernard Merkle walks to physiology class on a snowy day, does his brain signal the blood vessels in his skin to dilate or to constrict. Why?

CLINICAL TERMS AND DISEASES OF THE INTEGUMENTARY SYSTEM. Identify the following disorders of the skin by matching the disorder with its description. Write the correctly spelled name in the blanks. There may be a repetition of answers.

acne vulgaris
boil
burn
carbuncle
dermatitis
Herpes simplex

Herpes zoster
pediculosis
psoriasis
ringworm
tumors
wart

_____ 93. pimples

_____ 94. immunological disease caused by a virus

_____ 95. an inflammation of the dermis

_____ 96. abscessed hair follicle

_____ 97. "tinea", fungal infection of the skin

_____ 98. elevated skin caused by viral penetration

_____ 99. inherited raised flaky skin condition

_____ 100. a growth of useless cells, sometimes serious

_____ 101. shingles

_____ 102. lice are associated with this immunological disease

_____ 103. a "staph infection"

_____ 104. a mass of boils

_____ 105. denaturing of proteins in body tissue by fire, chemicals, radiation, or electricity

June 1994

Dear Doctor,

For each of the following diseases of the integumentary system, select a treatment from the list provided. Place the letter of the treatment in the blanks. Thank you.

Nurse Nightingale

____ 106. Acne vulgaris

____ 107. first degree burn

____ 108. third degree burn

____ 109. furuncle

____ 110. eczema

____ 111. Herpes simplex type 1

____ 112. Herpes simplex type 2

____ 113. Herpes zoster

____ 114. crabs

____ 115. psoriasis

____ 116. "jock itch"

____ 117. "Athlete's foot"

____ 118. freckles

a. over-the-counter tinea fungicide

b. benzyl benzoate and frequent cleansing of affected area

c. acyclovir into lesions and pain relievers to relieve the symptoms

d. no treatment is prescribed

e. insecticide lotion or dip

f. corticosteroid ointments, proper diet, and stress management

g. pain relievers to mask the symptoms.

h. hydrocortisone ointment to reduce inflammation

i. antibiotics

j. sunburn spray for comfort

k. skin grafting

STUDY SUGGESTION. When you have worked through this study guide for Chapter Six, read the Chapter Summary in your textbook aloud. Read as if you were a teacher giving a lecture. Try to keep your imaginary students awake by reading with enthusiasm. Hearing a meaningful lesson spoken with authority makes an impression on your own memory....and that's not imaginary!

LABELS AND LISTS

1. List five functions of the integumentary system.

A.

B.

C.

D.

E.

2. List the two distinct layers of the skin.

A.

B.

3. List the five layers of the epidermis from the deepest layer to the most superficial.

A.

B.

C.

D.

E.

4. List three frequently encountered types of skin cancers.

A.

B.

C.

5. Label the diagram of a hair follicle.

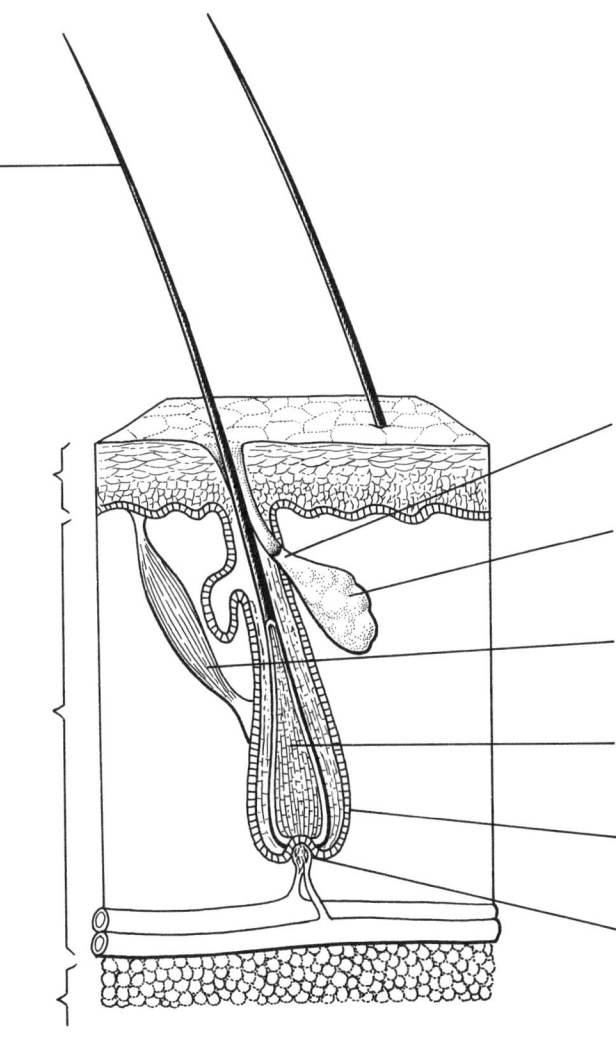

6. Label the diagram of a nail.

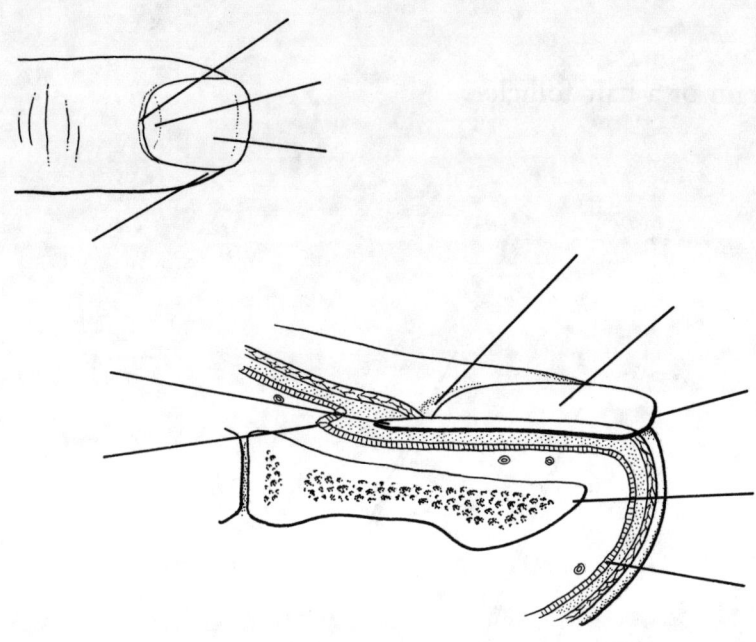

ADDITIONAL STUDY

Read the Chapter Summary (pages 149 - 150) out loud. Write down the definitions of all the KEY TERMS (page 150).

Review the figures in your textbook and understand the answers to the questions associated with each one. The answers are on pages 151.

Having studied this chapter, close your book, put away your notes, and test yourself by **writing** the answers to the "CONCEPTS CHECKS" and "QUESTIONS FOR REVIEW" in your text (pages 138, 140, 143, 147, 150). Prove to yourself that you can confidently answer these questions now, and you will be more likely to have confidence during the exam.

The day before an exam, read the "Learning Objectives", page 134, and review any of the sections which you think will cause you a problem.

CHAPTER 7: THE SKELETAL SYSTEM

In this chapter, you will distinguish between types of bones, see what is inside bones, describe bone composition and growth, identify the 206 bones of the skeleton, and study types of joints and their movements. Set the scope of your study by reading the chapter outline and learning objectives in your textbook. This is a great chapter!

SUGGESTION. Make your own vocabulary list as you read each section of the chapter. It will help keep your attention. Additionally, write out the answers to the questions accompanying the art work and those in the boxes at the end of sections of text. Writing the answers will help you recall them much more effectively than just "thinking" the answer! Check your answers. Remember, just moving your eyes over words is not the kind of "reading" that makes for true learning.

CONTENT MASTERY

BONE STRUCTURE. Read this section in your text. Read it aloud or "mutter" as you read to help focus your attention. Match each of the terms with the correct definition or statement. Write the word in the blank. Pay attention to spelling. A term may be used more than once, or not at all.

_____ 1. the thigh bone, for example

_____ 2. the sternum, for example

_____ 3. the vertebrae, for example

_____ 4. the bones of the wrist, for example

_____ 5. the long central shaft of a bone

_____ 6. thin layer of cartilage on the articulating surfaces of bone

_____ 7. another word for articulation

_____ 8. extreme ends of bone

_____ 9. connective tissue on the diaphysis

_____ 10. type of densely packed bone tissue

articular cartilage
compact bone
diaphysis
endosteum
epiphysis
flat bone
irregular bone
joint
lamellae
long bone
medullary cavity
osteoblast
osteoclasts
osteocytes
osteon
osteonic canals
periosteum
red marrow
short bone
spongy bone
Volkmann's canal
yellow marrow

_____ 11. type of loosely packed bone tissue found in the epiphyses

_____ 12. location of yellow marrow in a long bone

_____ 13. the youthful stage in the life of a bone cell

_____ 14. mature bone cells

_____ 15. wandering bone cells which dissolve minerals

_____ 16. bone cell which produces matrix but is not trapped by it

_____ 17. bone cell trapped in a lacuna surrounded by matrix

_____ 18. structure composed of osteonic canal, lamellae, osteocytes, and canaliculi

_____ 19. canals in the bone matrix through which blood vessels of the osteons interconnect

_____ 20. site of hematopoiesis (blood cell formation)

articular cartilage
compact bone
diaphysis
endosteum
epiphysis
flat bone
irregular bone
joint
lamellae
long bone
medullary cavity
osteoblast
osteoclasts
osteocytes
osteon
osteonic canals
periosteum
red marrow
short bone
spongy bone
Volkmann's canal
yellow marrow

BONE DEVELOPMENT, GROWTH, AND REMODELING. Complete the following statements as you read concerning bone development in your text.

21. Bones develop from two types of embryonic tissue: from _____ and from _____. After birth, bones grow in two directions, _____ and _____. Throughout adulthood, bone remodeling continues in order to _____.

22. When does bone development begin? _____.

23. The word "intramembranous" literally means _____.

24. The word "endochondral" literally means _____.

25. In what way are these two types of bones alike? _____.

26. Define ossification. _____.

27. The function of osteoblasts is _____.

BONE DEVELOPMENT, GROWTH, AND REMODELING. Compare and contrast intramembranous and endochondral bones by completing the following chart.

	INTRAMEMBRANOUS	ENDOCHONDRIAL
What secretes bone matrix?	28a.	b.
When does development begin?	c.	d.
Describe development	e.	f.
Examples of bones formed	g.	h.

101

Match the following phrases with the words in the right column. Place the correctly spelled word in the blank provided. Choices may be used more than once, or not at all.

_____ 29. Cartilage- producing cells

_____ 30. Hyaline cartilage producers

_____ 31. Location of primary ossification center

_____ 32. Location of secondary ossification center

_____ 33. Epiphyses-to- diaphyses cartilage

_____ 34. At end of epiphysis, where joint forms

_____ 35. Lengthwise expansion of endochondral bone

_____ 36. Where thigh bone, for example, grows longer

_____ 37. Line of ossification between epiphysis and diaphysis

_____ 38. Femur's growth in width

_____ 39. Cells responsible for enlargement of medullary cavity by destruction of surrounding bone

**appositional growth
articular cartilage
center of diaphysis
chondroblasts
epiphyseal line
epiphyseal plate
epiphysis
interstitial growth
osteoclasts**

TRUE/FALSE. Mark the following statements as "true" or "false." Rewrite each false statement as a true statement in the space provided by changing the italicized word(s).

_____40. After full height is achieved, the *epiphyseal plate* disappears.

_____41. In old age, bones *become inactive*.

_____42. Osteoblasts reabsorb bone in later life *as well as in youth*.

_____43. Bone remodeling occurs *mainly* during the repair of breaks and fractures.

_____44. In general, the bones that receive *more stress*, such as the femur, undergo more frequent remodeling than the bones of the breastbone.

_____45. Excess minerals in the blood are stored in *new* bone tissue.

_____46. In a green stick fracture, the break extends *through the bone*.

_____47. Inflammation and bleeding following a break in a bone are a *dangerous* sign.

_____48. A mass of protein fibers, produced after bone injury, is called a *prothallus*.

_____49. Several weeks after a fracture, the mass of new bone is called the *osseous callus*.

_____50. Another term for a closed fracture is *"green stick"*.

ORGANIZATION OF THE SKELETON. The majority of your study will involve labeling the skeleton. First, memorize the names of the bones as you read your text.

51. Two major parts of the skeleton are the _____ and _____ skeleton.

52. How many bones are there in the human body? _____

53. The structural features of bones relate to their functions. True/False

54. Study table 7-1. Cover the description and write out the correct term from memory. Go through the table again covering the terms. Locate the examples on the figures.

Match the following phrases with the choices. They may be used more than once, or not at all. For more of a challenge, fill in the blanks without consulting the list of choices!

_____	55. Number of bones in the skull	3 - 5
		5
_____	56. Rigid, narrow joints in the skull	7
		12
_____	57. Air-filled chambers of the skull	22
		30
_____	58. Inflammation of sinus mucous membranes	alveolar process
		atlas
		axis
_____	59. Forehead bone	cleft palate
		external auditory
_____	60. Another word for eye socket	meatus
		foramen
_____	61. Top, lateral bones of skull	magnum
		frontal
_____	62. Hole in occipital for spinal cord	hyoid
		mandible
_____	63. Name of the first vertebra	maxillary sinuses
		nasal septum
_____	64. Hole leading to inner parts of ear	orbits
		parietals
_____	65. Cheekbone	sella turcica
		sinuses
_____	66. Supports tongue	sinusitis
		styloid process
_____	67. "Turkish saddle" -shaped process	sutures
		turbinates
_____	68. Houses the pituitary gland	zygomatic
		zygomatic arch

_____ 69. Largest of the sinuses   3 - 5
_____ 70. Form tooth sockets   5
_____ 71. Separation of maxillary bones   7
_____ 72. Divides nasal cavity into two   12
                              portions   22
_____ 73. The nasal conchae   30
_____ 74. The only movable bone of the   **alveolar process**
                              skull   **atlas**
_____ 75. The nonarticulating bone   **axis**
_____ 76. Number of cervical vertebrae   **cleft palate**
_____ 77. Number of thoracic vertebrae   **external auditory**
_____ 78. Number of lumbar vertebrae   ** meatus**
_____ 79. Number of sacral vertebrae in   **foramen**
                              sacrum   **magnum**
_____ 80. Number of coccygeal vertebrae   **frontal**
                              in coccyx   **hyoid**
                                                                  mandible
                                                                  maxillary sinuses
                                                                  nasal septum
                                                                  orbits
                                                                  parietals
                                                                  sella turcica
                                                                  sinuses
                                                                  sinusitis
                                                                  styloid process
                                                                  sutures
                                                                  turbinates
                                                                  zygomatic
                                                                  zygomatic arch

TRUE/FALSE. Mark the following statements as "true" or "false." Rewrite each false statement as a true statement in the space provided by changing the italicized word(s).

_____81. The twelve thoracic vertebrae are *smaller* than the cervical vertebrae.

_____82. The only vertebrae that articulate with ribs are *cervical* vertebrae.

_____83. Thoracic vertebrae are unique in regards to their *facets*.

_____84. The lumbar vertebrae are thicker than the thoracic vertebrae due to their need for *flexion*.

_____85. An exposed spinal cord is *sometimes* seen in spina bifida cases.

_____86. Swayback is also known as *kyphosis*.

_____87. A lateral bend of the spine is called *scoliosis*.

_____88. Severe lateral curve of the spine can be *successfully* treated with electrical stimulation of selected muscles.

_____89. Men have *23* ribs while women have 24.

_____90. False ribs are those which *do not articulate with any other bones* and are also known as floating ribs.

_____91. There are *15* phalanges on each hand.

_____92. The rough surface of the *ischial spine* supports your body weight while sitting.

_____93. The largest foramen of the skeleton is the *foramen magnum*.

ARTICULATIONS. Fill in the following chart as you read.

TYPE OF JOINT BASED ON MATERIAL	STRUCTURE	MOVEMENT(S) ALLOWED	EXAMPLE
94.			
95.			
96.			

Match the terms. Some descriptions require more that one term.

_____ 97. "soft spots"

_____ 98. Distal joint between tibia and fibula

_____ 99. Tooth-to-socket

_____ 100. Joint with hyaline cartilage

_____ 101. Joint with fibrocartilage

_____ 102. Damaged disc

_____ 103. Removal of part of the vertebra

bursa
cartilaginous joint
fontanels
gomphosis
herniated disc
laminectomy
suture
syndesmosis
synovial joint
synovial cavity
tendon sheath

_____ 104. Joint allowing most movement

_____ 105. Fluid-filled space between

_____ 106. Between ribs and sternum

_____ 107. Between cranial bones

_____ 108. Flat sacs of synovial fluid to reduce friction

_____ 109. Long sacs of synovial fluid

bursa
cartilaginous joint
fontanels
gomphosis
herniated disc
laminectomy
suture
syndesmosis
synovial joint
synovial cavity
tendon sheath

Read about synovial joints before you fill in the following section. Then write the name of one of the six synovial joints that is most closely described by the following statements and examples.

gliding, hinge, pivot, condyloid, saddle, ball and socket

_____ 110. Two flat surfaces which are approximately the same in size

_____ 111. Round head fitting into a cup-shaped depression

_____ 112. Convex surface applied to a concave surface as in the knee

_____ 113. Cylindrical process that rotates within a ring

_____ 114. Occurring between trapezium and metacarpal of the thumb

_____ 115. Oval-shaped process of one bone fitting into a cavity of another

_____ 116. Found between breastbone and clavicle

_____ 117. Found between axis and atlas

_____ 118. Hip joint, for example

An understanding of movements is essential to understand muscles in the next chapter. Read about the types of movement allowed by various synovial joints and then fill in the following blanks.

_____ 119. decrease in angle between bones

_____ 120. top of the foot moves toward the ankle

_____ 121. increase in angle between bones

_____ 122. an increase in angle beyond anatomical position

_____ 123. standing on the toes

_____ 124. movement away from the midline

_____ 125. movement toward the midline

_____ 126. distal end of a bone makes a circle while proximal end remains relatively stationary

_____ 127. movement around an axis

_____ 128. palm turns posteriorly

_____ 129. palm turns anteriorly

_____ 130. sole points laterally

_____ 131. sole points medially

_____ 132. body part moves forward, parallel to the ground

_____ 133. returning the body part to anatomical position (after #130)

abduction
adduction
circumduction
dorsiflexion
eversion
extension
flexion
hyperextension
inversion
plantar flexion
pronation
protraction
retraction
rotation
supination

Consider the movements of a baseball pitcher during a game. Write the physiological term which most accurately describes his actions.

_____ 134. The pitcher first looks at first and third bases.

_____ 135. Next he looks straight up at a passing blimp.

_____ 136. Then he looks straight down at a snake in the grass.

_____ 137. He looks up from the snake at the catcher.

_____ 138. He signals his team by raising his arm laterally.

_____ 139. Of course he brings it back .

_____ 140. He signals the catcher by moving his jaw forward, parallel to the ground.

_____ 141. Finally, he pitches the ball moving his outstretched arm in a circular fashion.

STUDY HINT. Abduction is alphabetically before adduction. Likewise, abduction must occur before adduction can. Supination involves turning the hand anteriorly or upward. Supination is an action which you use to eat supper.

CLINICAL TERMS. Match the following clinical terms of the skeletal system with their descriptions given.

_____	142. Vitamin D could prevent this disease	**achondroplasia**
		gout
		osteitis deformans
_____	143. Most common form of arthritis	**osteoarthritis**
		osteomalacia
		osteomyelitis
_____	144. Accumulation of uric acid in joints	**osteoporosis**
		rheumatoid arthritis
		rickets
_____	145. Inherited disease resulting in dwarfism	**tumors of bone**

_____ 146. Also called Paget's disease

_____ 147. In children, weakened bones, bowed limbs, soft skull

_____ 148. Joint becomes completely fused with bone, autoimmune disease

_____ 496. Rare, painful, often life-threatening

_____ 150. Imbalance between bone formation and bone resorption

_____ 151. Caused by *Staphylococcus aureus* infection

_____ 152. Thinning and softening of bones in adults, skeletal deformities

_____ 153. Seen in postmenopausal women and elderly men

_____ 154. Treated with antibiotics

LABELS AND LISTS

1. List five functions of the skeletal system.

    a.   b.   c.

    d.   e.

2. Name the four basic types of bones.

    a.   b.   c.

    d.

3. What are the three types of bone cells in living bone?

    a.   b.   c.

4. Fetal bone development can be described by the manner in which the bones cells arise. Name the two types of bone development.

    a.   b.

5. Name three bones that develop in the fetus between thin sheets of embryonic membrane.

    a.   b.   c.

6. Name three bones that develop from cartilage.

    a.   b.   c.

7. Name the three types joints based upon their movement.

    a.   b.   c.

8. List the two mineral salts which combine to form hydroxyapatite

   a.   b.

9. Label the parts of a long bone.

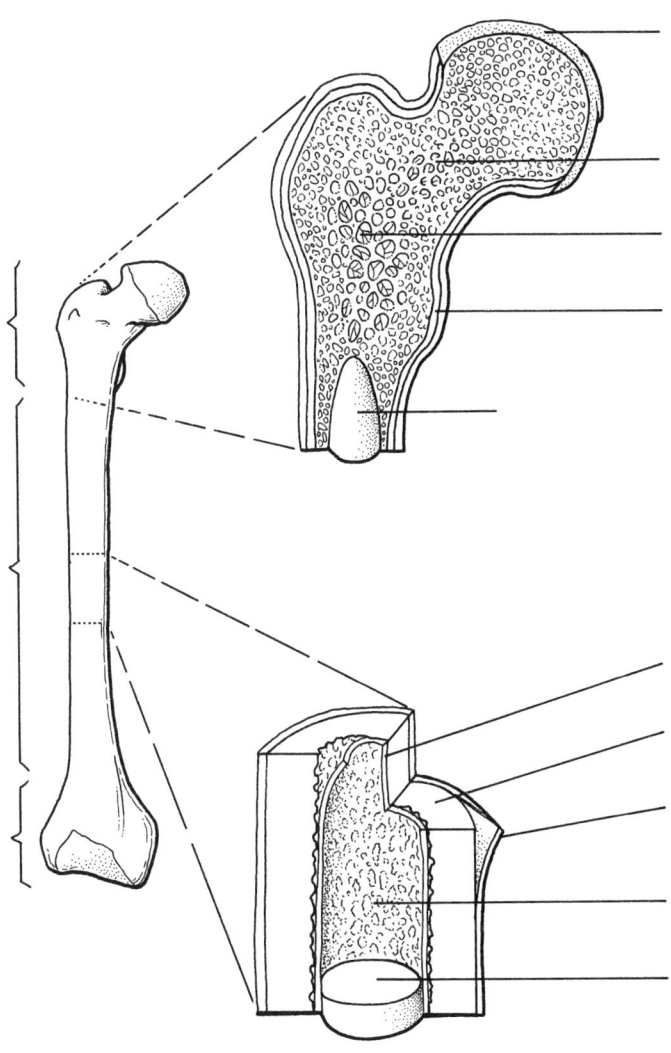

10. Label the microscopic section of bone.

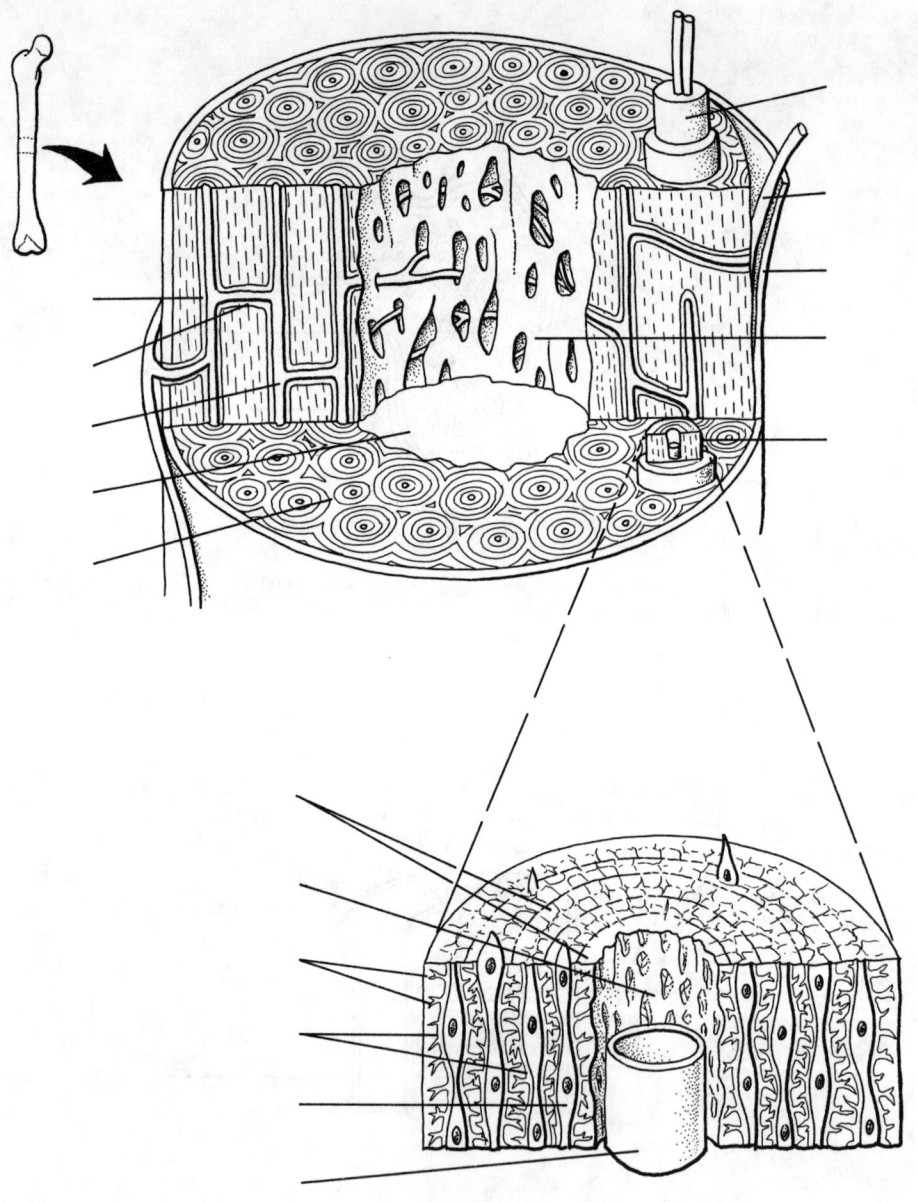

11. Label the skeleton.

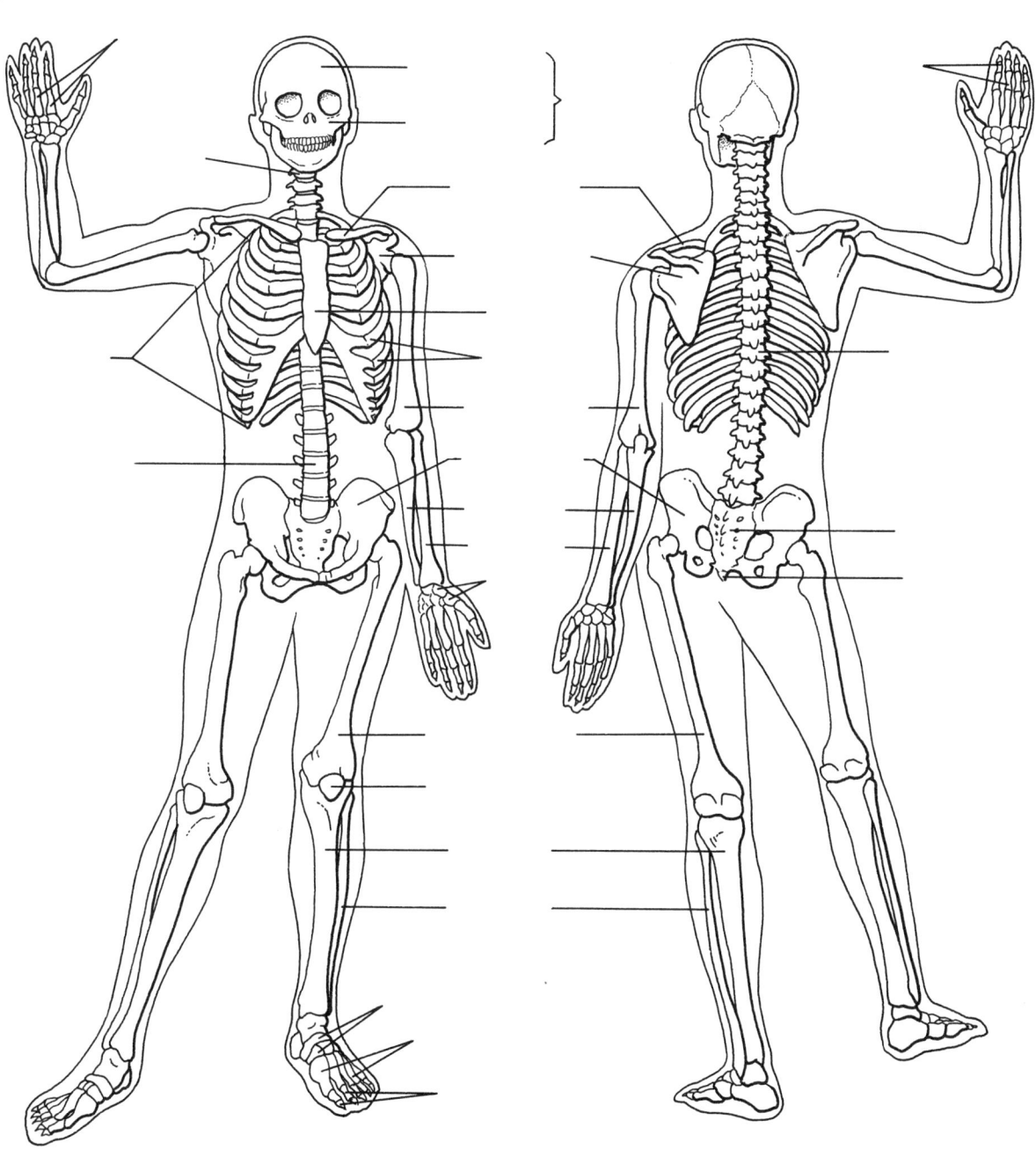

12. List the four components of the axial skeleton.

    a.   b.   c.

    d.

13. List the four components of the appendicular skeleton.

    a.   b.   c.

    d.

14. List the surface feature of bones described by each of the following:

    _____ a depression or groove

    _____ an opening or hole

    _____ a narrow opening

    _____ a large rounded protrusion

    _____ a small rounded protrusion

    _____ a large blunt process on the femur

    _____ a usually rough, elevated area

    _____ a smooth articular surface

    _____ any projection from the surface

    _____ a narrow or pointed projection

15. What are three functions of sinuses?

    a.

    b.

    c.

16. List five bones of the skull that contain sinuses.

    a.   b.   c.

    d.

17. List six aspects of the male pelvic girdle that are different from the female.

    a.   b.

    c.   d.

    e.   f.

18. Name the three coxal bones in a newborn infant.

    a.   b.   c.

19. List the bones of the upper extremity in order, proximal to distal.

    a.   b.   c.

    d.   e.   f.

20. Joints may be classified on the basis of the tissues that binds them. Name three types.

    a.   b.   c.

21. List 6 types of synovial joints.

    a.   b.   c.

    d.   e.   f.

22. Label the actions illustrated in the following photographs.

 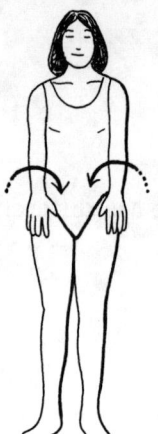 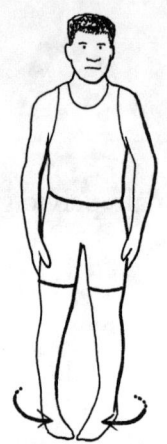 

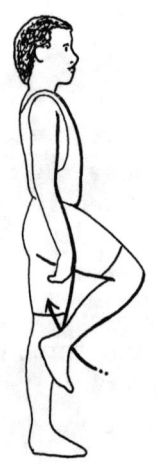 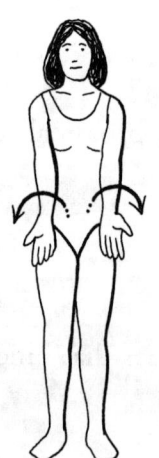  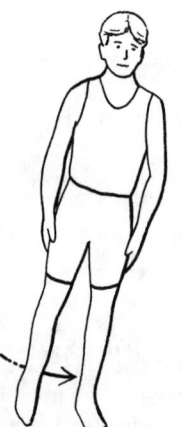

23. List four homeostatic functions of the skeletal system.

   a.   b.   c.

   d.

Practice labelling all of the surface features of each bone as shown in your textbook.

TEST YOURSELF

MULTIPLE CHOICE: Place the letter corresponding to the best answer in the space provided.

_____ 1. Which of the following is NOT a component of bone tissue?
   a. calcium phosphate
   b. collagen
   c. osteocytes
   d. hydroxyapatite
   e. Keratohyalin

_____ 2. In a cross section of bone, which of the following cell types would be most likely seen on the surface of the bone?
   a. osteoblasts
   b. osteocytes
   c. osteoclasts
   d. endosteum
   e. periosteum

_____ 3. Which of the following substances is most prevalent in a sample of bone?
   a. collagen
   b. hydroxyapatite crystals
   c. osteocytes
   d. blood
   e. nerve tissue

_____ 4. Which statement best describes spongy bone?
   a. Spongy bone contains a dense matrix.
   b. Spongy bone contains more spaces filled with red marrow than compact bone.
   c. Spongy bone is found particularly in the diaphysis of long bones.
   d. Canaliculi are not necessary in spongy bone because of the thin trabeculae.
   e. Both b and d accurately describe spongy bone.

_____ 5. Osteocytes (bone cells) receive nourishment in compact bone by way of
   a. Blood vessels that pass through osteonic canals.
   b. Blood vessels formed in the red marrow and passing through short canaliculi.
   c. Diffusion of oxygen through the matrix
   d. Osteoclasts which wander throughout bone tissue under the influence of hormones.

_____6. The smooth surfaces on thoracic vertebrae which serve to articulate with the ribs are called
   a. superior articulating processes
   b. laminae
   c. pedicles
   d. facets
   e. thoracic processes

_____7. Which of the following is NOT found in the appendicular skeleton?
   a. phalanges
   b. tarsals
   c. thoracic cage
   d. coxal bones
   e. humerus

_____8. A scientist studies a bone which forms by the replacement of a cartilaginous model by bone tissue. This type of bone is called
   a. an endochondral
   b. epiphyseal
   c. interstitial
   d. chondroblast
   e. intramembranous

_____9. The shaft of a long bone is the
   a. diaphysis
   b. epiphysis
   c. periosteum
   d. epiphyseal plate
   e. medullary cavity

_____10. A bone which has achieved its final adult size no longer contains living bone cells.
   a. true
   b. false

_____11. Which bone does not articulate with any other?
   a. sphenoid
   b. floating rib
   c. patella
   d. hyoid
   e. mandible

_____12. A student observes a hole through the sphenoid. In general this hole is a
   a. fossa
   b. foramen
   c. tubercle
   d. fissure
   e. facet

_____13. The lumber vertebrae
   a. are five in number
   b. have spinous processes
   c. are larger and thicker than other vertebrae
   d. support the weight of the upper body
   e. all of the above

____14. An archeologist discovers a skeleton in a cave in the Near East. He determines that it is the skeleton of a male. Which of the following features most probably led him to this determination?
a. height of the skeleton
b. shape of pelvic inlet
c. number of ribs
d. the number of vertebrae
e. all of the above

____15. The bone of the foot which articulates with the bones of the leg is
a. talus
b. fibula
c. tarsal
d. calcaneus
e. malleolus

____16. Which of the following joints is the most moveable?
a. suture
b. syndesmosis
c. gomphosis
d. cartilaginous
e. synovial

____17. The skeleton helps maintain homeostasis of the body by
a. providing support
b. being the site of hematopoiesis
c. storing minerals
d. protection of soft organs
e. all of the above

____18. Bending the knee is an example of
a. pronation
b. adduction
c. flexion
d. retraction
e. extension

____19. The hard, outer portion of a bone is composed of
a. spongy bone
b. red marrow
c. articular cartilage
d. compact bone
e. epiphyses

____20. Which of the following is necessary for the bone of a child to continue to grow in length?
a. active chondroblasts in the epiphyseal plate
b. visible ossification in the epiphyseal line
c. synovial fluid production
d. appositional growth and remodelling
e. a living periosteum

____21. Which of these is an important aspect of bone remodelling?
   a. osteoblastic deposition and osteoclastic removal of bone
   b. storage of calcium
   c. healing of a bone fracture
   d. returning stored phosphate to the blood
   e. all of the above

____22. Sinuses
   a. are a real pain
   b. contain synovial fluid
   c. function to reduce the weight of the skull and resonate the voice
   d. are located in the occipital
   e. are part of the nasal cavity

____23. The distal portion of the nose is composed of
   a. nasal bones
   b. inferior nasal conchae
   c. perpendicular plate of the ethmoid
   d. cartilage
   e. vomer

____24. A "slipped disc" is
   a. the movement of a computer diskette behind a desk
   b. a damaged intervertebral disc
   c. the lateral movement of vertebrae in scoliosis
   d. a knee injury

____25. A transverse foramen
   a. permits passage of an artery to the brain
   b. provides a site of muscle attachment
   c. allows rotation of the head
   d. is an opening in the occipital bone
   e. extends laterally from the body of a thoracic vertebra

____26. A compound fracture
   a. is the same thing as a closed fracture
   b. involves breaking the bone into more than one piece
   c. is also called "comminuted"
   d. occur when the skin is pierced by the bone
   e. is usually incomplete

____27. The uppermost vertebra
    a. is the atlas
    b. has no body
    c. has a transverse foramen
    d. is cervical
    e. all of the above

____28. Which of the following does not attach to the sternum?
    a. hyoid
    b. costal cartilage
    c. rib pair four
    d. clavicle
    e. true rib

____29. Synovial joints
    a. are moveable
    b. contain synovial fluid
    c. may have bursae
    d. include gliding joints
    e. all of the above

____30. An insertion is
    a. the point of attachment of a muscle tendon to a more moveable bone
    b. an inclusion inside a synovial joint
    c. a type of fracture involving many small bone fragments
    d. a type of ,movement found at some synovial joints
    e. the contents of the medullary cavity

Answer all of the questions in the review and those accompanying the figures in your textbook.

THE SKELETAL SYSTEM
Robert W. Bauman, Jr., Ph.D.

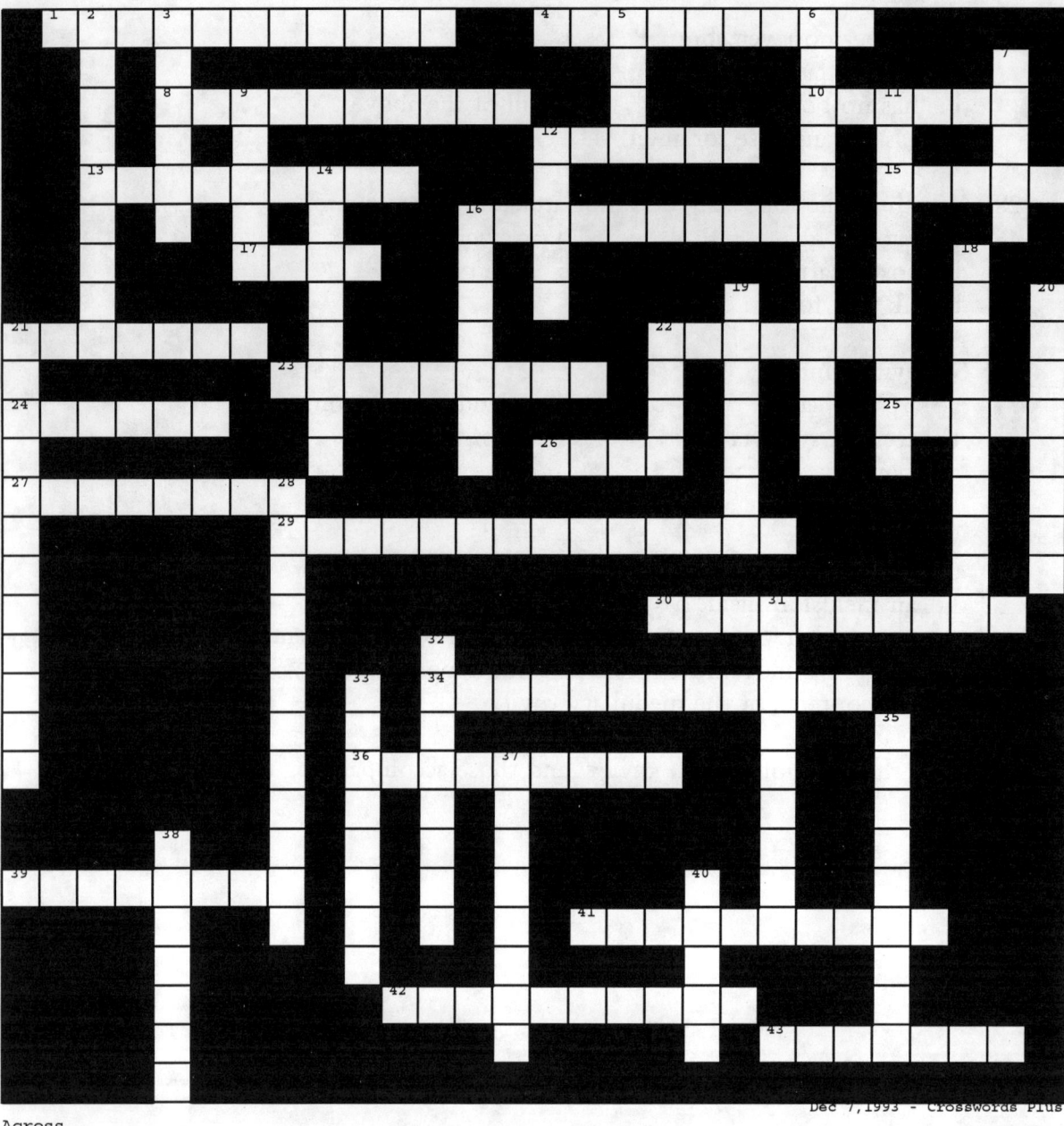

Across

1. recycling of bone
4. movement toward the midline
8. bony plate
10. bone containing red marrow
12. inferior vertebral curve
13. lateral bend of vertebrae
15. lower leg bone
16. small channels
17. greater length than width
21. large arm bone
22. forehead
23. shaft of bone
24. content of bone
25. bone which is same width and length
26. arthritis with uric acid crystals
27. twelve vertebrae in this group
29. bone crystal
30. rough surface feature
34. joint
36. cavity in diaphysis
39. exaggerated thoracic curve
41. incomplete fracture
42. bone covering
43. concentric ring of osteon

Down

2. lining of medullary cavity
3. haversian system
5. fibrocartilage between vertebrae
6. creation of bone
7. non-articulating bone
9. along the axis
11. bone builder
12. cavity in skull bone
14. moveable joint
16. rounded prominence
18. matrix-dissolving cell
19. tightly packed bone
20. soft spot
21. blood formation
22. thin bone
28. cartilage cell
31. end of long bone
32. jaw
33. fracture which breaks the skin
35. bone cell
37. swayback
38. hole in bone
40. longest human bone

Dec 7, 1993 - Crosswords Plus

CHAPTER 8: THE MUSCULAR SYSTEM

The muscular system is "in" for the 90's! It's "rad" to rattle off names of muscles, and it is common to flip through the cable channels and see muscle-bound men and women flexing to delineate muscle groups. Aside from the show, muscles are distinguished by their contractility, excitability, extensibility, and elasticity. You will find the descriptions of how muscles work (their physiology) to be fascinating, and surely you will find that the anatomy of the muscular system is poetry in motion! This is a great chapter!

CONTENT MASTERY

MUSCLE STRUCTURE. After reading both the introduction and the first section of material, fill in the blanks with the best answer from the text.

_____ 1. The ability of a muscle cell to shorten in length is called...

_____ 2. The ability to receive and respond to a stimulus is ...

_____ 3. The ability of a muscle cell to stretch is called...

_____ 4. The ability of a cell to return to its resting form after stretching is called..

_____ 5. The type of muscle tissue found in the muscular system is...

_____ 6. The connective tissue that binds a muscle and provides a route for blood vessels and nerves to travel is called...

_____ 7. The type of fascia that surrounds a muscle is called ... fascia.

_____ 8. The outermost covering of the entire muscle is the ..., which is deep fascia.

_____ 9. A thick band of dense connective tissue that forms most connections between muscle and bone is called a ...

_____ 10. A broad sheet of dense connective tissue that may attach a muscle to a bone or to another muscle is an ...

_____ 11. A tendon rupture is a serious injury calling for ... to reconnect the torn ends.

_____ 12. Overuse of a tendon, called ..., results in pain and inflammation.

_____ 13. A single cell of skeletal muscle tissue is called a ...

_____ 14. The plasma membrane of a muscle cell is called a...

_____ 15. The cytoplasm of a muscle cell is called...

_____ 16. Most of the energy for the contraction of a muscle cell is furnished by the ATP of the ...

_____ 17. Calcium storage occurs at the .... reticulum which is similar to the endoplasmic reticulum of other cells.

_____ 18. Between adjacent sarcoplasmic reticular sacs is a tube called the ... tubule which unites with the sarcolemma.

_____ 19. ... are tiny parallel fibers which extend the length of a muscle fiber.

_____ 20. There are two types of protein filaments in myofibrils: thick filaments and ...

_____ 21. Thick filaments are composed of ...

_____ 22. Thin filaments are composed of ..., troponin, and tropomyosin.

_____ 23. The striations seen on a muscle cell are made of dark bands called ... and light bands called I bands.

_____ 24. The segment of a myofibril between two Z lines is called a ...

_____ 25. The difference in charges (+ outside and - inside) across a plasma membrane causes a small voltage difference which is called ...

_____ 26. The end of a motor neuron, a motor end plate, and the narrow space in between is referred to as the ...

_____ 27. Acetylcholine is located in tiny sacs called ... which are located at the terminal end of motor neurons.

_____ 28. When a nerve or muscle cell is stimulated, there is a brief reversal of charges across the cell membrane. This reversal is called an ...

_____ 29. The functional unit consisting of a single motor neuron and the many muscle fibers it stimulates is called a ...

_____ 30. ... is a term for a chemical that carries a signal from one nerve end to either another neuron or to a muscle cell.

_____ 31. "ACh" is an abbreviation for ...

THE PHYSIOLOGY OF MUSCLE CONTRACTION.

_____32. Which of the following is most accurate concerning the "at rest" muscle fiber?
   a. Calcium ions are traveling to the muscle fiber to be used during contraction.
   b. ATP is chemically attached to myosin proteins.
   c. The thin filaments are ready with actin, troponin, and tropomyosin.
   d. Both "b" and "c" are correct statements, but "a" is incorrect.

_____33. What is the chemical stimulus that begins a muscle contraction?
   a. actin release
   b. ACh release
   c. calcium ions are released into the sarcoplasm
   d. troponin release

_____34. Muscle contraction involves calcium ions, myofibrils, ATP, myosin, and troponin. After reading the two paragraphs about muscle contraction, select the false statement from those below.
   a. Calcium binds to troponin molecules which reshapes the actin and troponin molecules.
   b. Connections are made between thin and thick filaments.
   c. Calcium (with the catalyzing action of myosin) causes the release of potassium and energy from ATP molecules.
   d. All of the energy released in the breakdown of ATP increases body temperature (which is why the body becomes warm during exercise).

35. Place the following events in muscle contraction in the correct sequence.

____ ____ ____ ____

   a. Energy released in the breakdown of ATP increases body temperature (which is why the body becomes warm during exercise).
   b. Calcium binds to troponin molecules reshaping the actin and troponin molecules.
   c. Calcium (with the catalyzing action of myosin) causes the release of potassium and energy from APT molecules.
   d. Connections are made between thin and thick filaments.

36. Place the following events in muscle contraction in the correct sequence.

____ ____ ____ ____

   a. Z lines of the sarcomere are shortened as they are drawn together.
   b. Cross bridge movement
   c. Cross bridge formation
   d. Cross bridge release

_____37. Rigor mortis can be best described as
   a. muscular relaxation due to unavailability of ATP
   b. a condition of death when ATP collects in muscle tissue and causes total release of cross bridges and consequently muscular rigidity
   c. muscular rigidity due to unreleased cross bridges in the muscle cells
   d. a lack of ATP because of halted metabolic activities resulting in cold body temperatures

_____38. What biochemical change is most directly responsible for the return of a muscle fiber to a resting condition?
   a. absence of calcium ions
   b. the absence of acetylcholinesterase
   c. the release of acetylcholine
   d. the regeneration of ATP

_____39. Which of the following describes a direct use of the energy from ATP breakdown?
   a. the mechanical movement of cross bridges
   b. the breakage of cross bridge attachments from thin filaments
   c. the return of calcium to the sarcoplasmic reticulum
   d. all of the above

_____40. Energy for contraction is NOT provided from which of the following?
   a. ATP produced during the glucose-breakdown phase of cellular respiration
   b. creatine phosphate stored in muscle cells
   c. glucagon which is broken down to yield glucose to be used in cellular respiration to produce ATP
   d. fat molecules which contain the most potential energy

_____41. Which of the following is true of lactic acid?
   a. Lactic acid uses up the oxygen resulting in oxygen debt.
   b. Lactic acid is produced in muscle cells in the presence of oxygen
   c. Lactic acid accumulation results in soreness after exercise.
   d. Lactic acid is used during cellular respiration to synthesize AMP.

_____42. In muscles that are exercised strenuously for a prolonged period of time, the oxygen debt may lead to muscle fatigue.
   a. true
   b. false

_____43. Muscle fatigue is the inability of a muscle to contract normally.
   a. true
   b. false

_____44. When a muscle contracts in a spasm without relaxing, the result is
   a. a cramp
   b. oxygen debt
   c. muscle fatigue
   d. paralysis

_____45. As the diameter of a muscle fiber increases,
   a. the strength of its contraction decreases.
   b. the strength of its contraction increases.
   c. they are said to hypertrophy.
   d. both b and c above

_____ 46. Which of the following is NOT true of disuse atrophy?
   a. It is a condition in which muscle fibers increase in size.
   b. It is a condition in which muscle strength is decreased at a rate of 5% a day.
   c. It is a result of immobilization of muscles for an extended period.
   d. It could result in the replacement of muscle tissue with fibrous connective tissue.

SMOOTH MUSCLE AND CARDIAC MUSCLE. After reading the few paragraphs in your text, complete the following chart to compare the three types of muscle tissues.

	SKELETAL	SMOOTH	CARDIAC
47. Where is it found?	attached to skeleton		
48. How is it controlled			involuntary
49. Describe the shape of the fibers		spindle-shaped	
50. Are there striations?			
51. How would they place in a contraction speed race?			second place
52. How would they place in a strength competition?	first place		
53. Which can stay contracted the longest time?			

MUSCULAR RESPONSES. This section is full of potential essay questions! Read as if you were an instructor looking for good questions. Make a list of questions, fill in the blanks below, and then write answers for your own questions.

all-or-none response
complete tetanus
endurance
incomplete tetanus
isometric contraction
isotonic contraction
motor skill development

muscle tone
recruitment
subthreshold stimulus
threshold stimulus
treppe
twitch
wave summation

_____ 54. the weakest stimulus that can initiate a contraction

_____ 55. a stimulus that is too weak to cause a contraction

_____ 56. the complete contraction of a muscle fiber in response to a stimulus above the minimal amount

_____ 57. addition of motor units as stimulus strength increases

_____ 58. a rapid response to a single stimulus, a basic unit of muscle contraction

_____ 59. a muscle warm up phenomenon in which single twitches rapidly follow each other

_____ 60. a second, stronger contraction when a muscle receives a second stimulus before the first contraction cycle is complete

_____ 61. a continuous contraction due to a fusion of twitches

_____ 62. a type of tetanus in which only a small number of fibers contract affecting posture

_____ 63. the usual means of producing body movement

_____ 64. muscle tension, without shortening the muscle

_____ 65. learning a new coordination action

_____ 66. sustained muscular efforts with benefit to cardiovascular and respiratory systems

PRODUCTION OF MOVEMENT. One of your goals is to be able to define the following terms associated with movement. Though the definitions are given to you here in a matching form, increase the value of your study time by learning each definition. Be able to cover the answers in either column and give the correct match.

_____ 67. Define "origin."

_____ 68. Define "insertion"

_____ 69. Define "group action"

_____ 70. Define "prime movers."

_____ 71. Define "antagonists."

_____ 72. Define "synergists."

_____ 73. Define "fixators."

a. the coordinated response of a group of muscles that causes a body movement

b. the point of attachment to the more stationary bone

c. muscles in a group which cause the desired action

d. the point of attachment to the more movable bone

e. the muscles in a group which stabilize the origin of the prime mover

f. the muscles in a group that relax during the action

g. the muscles in a group which steady the movement

NAMING THE MAJOR MUSCLES OF THE BODY. Identify the literal translation of the Latin or Greek stems by writing the names of the major muscles of the body. Most of the stems are found in footnotes. Be sure to use correct spelling.

_____ 74. sphere of the eye

_____ 75. sphere of the mouth

_____ 76. trumpeter

_____ 77. cheek

_____ 78. one who chews

_____ 79. breast bone, clavicle, breast-resembling process

_____ 80. table-like

_____ 81. one who raises the shoulderblade

Buccinator
Levator scapulae
Masseter
Orbicularis oculi
Orbicularis oris
Pectoralis minor
Serratus
Sternocleidomastoid
Trapezius
Zygomaticus

		Biceps brachii
_____	82. saw	Deltoid
_____	83. of the breast, small	Gastrocnemius
		Gluteus maximus
_____	84. triangular	Gracilis
		Iliopsoas
_____	85. below the shoulderblade	Peroneus longus
		Quadriceps femoris
_____	86. above the spine	Rectus abdominis
		Sartorius
_____	87. rounded	Soleus
		Supraspinatus
_____	88. three heads in the arm	Subscapularis
		Tensor fascia latae
_____	89. two heads in the arm	Teres
		Triceps brachii

_____ 90. vertical presence of the abdomen

_____ 91. pertaining to the ilium and loin

_____ 92. great presence of the buttocks

_____ 93. slender

_____ 94. tenses fascia on the lateral

_____ 95. tailor

_____ 96. four heads of the femur

_____ 97. stomach of the lower leg

_____ 98. sole of the foot

_____ 99. long pin

Please note the spelling of these muscles. Practice them by copying each muscle name five times. Note also that "gastrocnemius" contains the letter "c"!

MAJOR MUSCLES OF THE BODY. The majority of your study time for the remainder of this chapter should be spent labeling the muscles on the diagrams in the "LABELS AND LISTS" section of this study guide. Make photocopies of the drawings before you label them so that you can label them again and again.

You should also note that the text provides tables of muscle groups with their origins, insertions, and actions. Be sure to ask your professor how much of this material is expected to be memorized. Each teacher emphasizes different areas and amount of material. Whatever is required for you to commit to memory, it would be advisable to cover the columns and quiz yourself. When you feel you have mastered the appropriate portions of the chart, make a blank chart to fill in without assistance from the text.

As an overview of other interesting information, fill in the following blanks.

100. The immediate treatment for pulled or torn muscles is a formula known as "RICE" which stands for ....

101. The "kissing" muscles are the ... and the ...

102. The largest muscle of the group which connects the pectoral girdle to the thorax is the ...

103. The muscle commonly known as the "lat" is the ...

104. The muscle known as the "swimmer's muscle" is the...

105. Injections are commonly given in a muscle called the ...

106. Evidence suggests that steroids are highly addictive. (true/false)

107. A vertical line (a ridge of connective tissue) extending from the sternum to the navel is called the ...

108. A muscle which is a common site for injections in the hip area is the ...

109. The ... muscles acquired their name from the butcher shop where their tendons were used to suspend pig meat during curing.

110. The gastrocnemius and the soleus insert via a common heel tendon known as the ...

HOMEOSTASIS. This short section lends itself beautifully to essay questions! Don't let the mention of the word "essay" produce anxiety in you. Learn to recognize the potential for discussion questions. Get a jump on the situation by creating (and answering) your own questions before your teacher can ask them. When you become competent with this study plan, you will rarely be surprised by essay questions!

111. The muscular system helps maintain homeostasis in what two ways?

112. How is body movement (provided by the muscular system) necessary for survival?

113. How does the muscular system affect the temperature regulation in the body?

CLINICAL TERMS OF THE MUSCULAR SYSTEM. Assume you are a nursing observer in a third world country when a local disaster absorbs the attention of all trained personnel. Due to your diligent study of human anatomy and physiology, you can direct the other less capable volunteers by solving the following problems.

_____ 114. Johnny the Jock is admitted with intense pain in his leg. He is very alarmed at the bluish color of his thigh. What should you tell him?
   a. "The Doctor is so busy, but we can schedule you for surgery tomorrow."
   b. "Do not worry. Bruising is common in Charley horses because of some muscle tearing. The bruise will go away."
   c. "The bluish color is due to lack of oxygen to the muscles because of a deficiency of acetylcholine. When someone can get to you, you can have an injection and notice the difference in a few hours."
   d. "The bruising is due to inflamed muscle tissue. It is probably a low-grade infection which can be cleared up quickly with antibiotics."

_____115. The patient in 115 is scheduled for an electromyography. She is asking what to expect. What should you tell her.
   a. "This procedure involves making a recording of the electrical activity in your muscles during contraction."
   b. "This procedure is like an x-ray and will be over in a second."
   c. "Electromyography is a shock treatment of some sort...like you see them do on television when they zap a crazy person."
   d. "This is a surgical procedure in which a small muscle is removed and stimulated electrically to cure neuromuscular problems. The muscle is then sutured back into the body."

_____116. A patient's chart notes chronic weakness, neuromuscular problems and acetylcholine deficiency. Which of the following is the disease?
   a. muscular dystrophy
   b. myalgia
   c. myasthenia gravis
   d. myositis

The patient's charts in rooms 117 through 122 have been shuffled around! Their medications are labeled by the name of their diseases. In order to be sure you have the correct medication for each, you have quizzed each patient as to their symptoms. Match the following patient's symptoms with the name of their disease.

Muscular Dystrophy   **Myotonia**
Myalgia   **Shinsplints**
Myoma   **Torticollis**
Myositis

_____ Room 117. Loss of strength progressively worsening, wrists showing signs of deformity

_____ Room 118. Uterine tumor. Benign.

_____ Room 119. Inflamed muscle following injury on football field.

_____ Room 120. An "uptight" feeling in the muscles of the back. Muscles not relaxing after contraction.

_____ Room 121. Young tennis player complaining of severe pain in anterior tibialis muscle.

_____ Room 122. Car accident victim. Patient's head is tilted to the right. Cannot straighten neck.

LABELS AND LISTS

1. List four properties of muscle tissue.

   a.

   b.

   c.

   d.

2. List three roles of muscles.

   a.

   b.

   c.

3. List three things required to shorten the sarcomeres of muscle fibers (that is, to cause a muscle contraction).

   a.

   b.

   c.

4. Label this drawing of the structure of a muscle fiber.

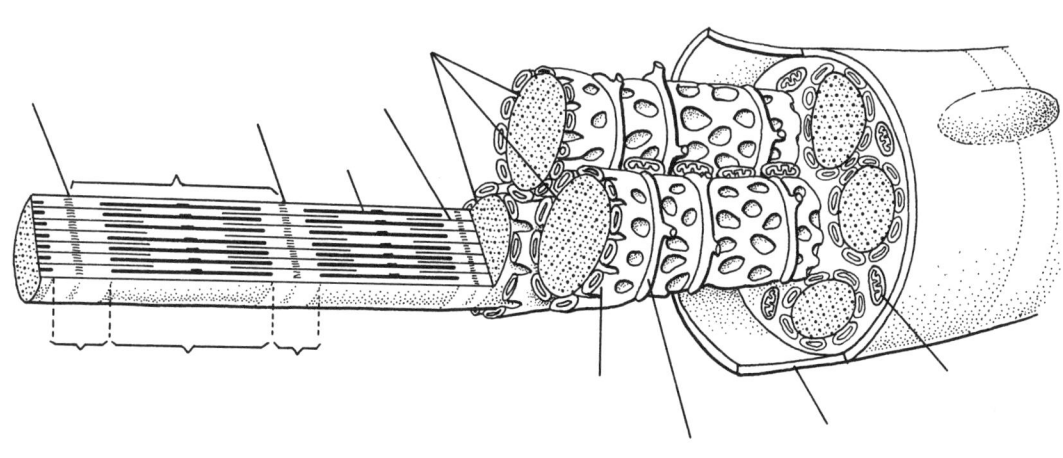

5. Label this drawing of the structure of a muscle.

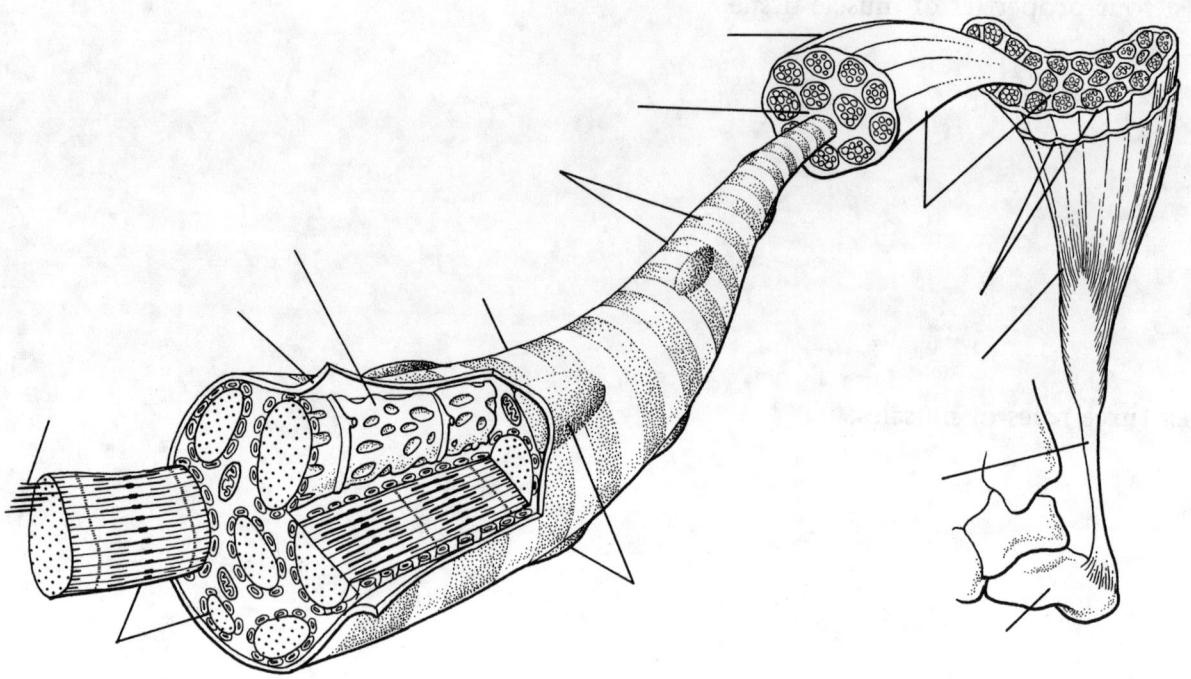

6. Label this drawing of a neuromuscular junction

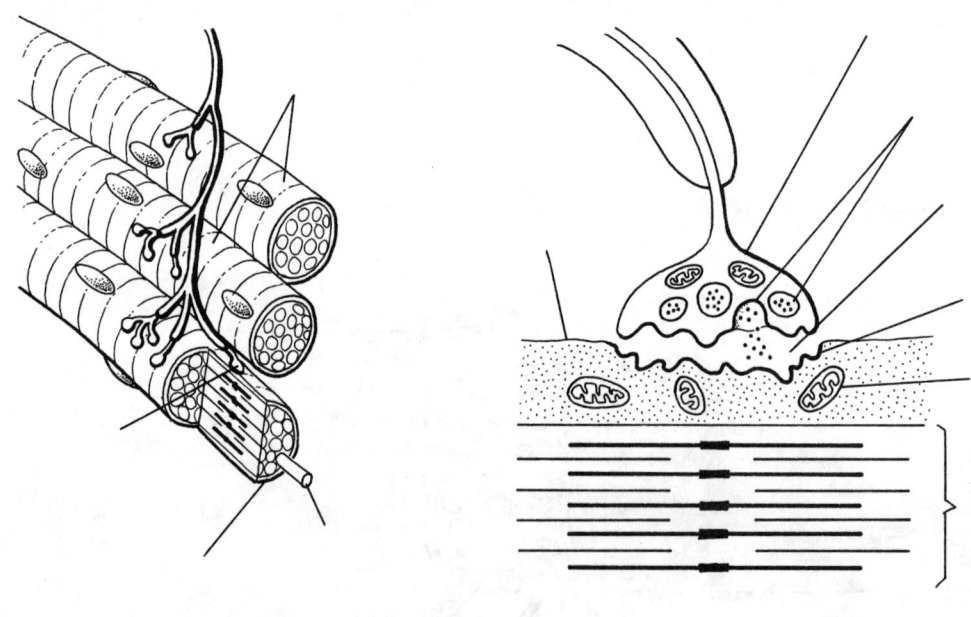

138

7. Label the muscles. What is the function of each muscle? Dependent upon the instructions of your teacher, also list the origin and insertion for each muscle on this and all subsequent figures.

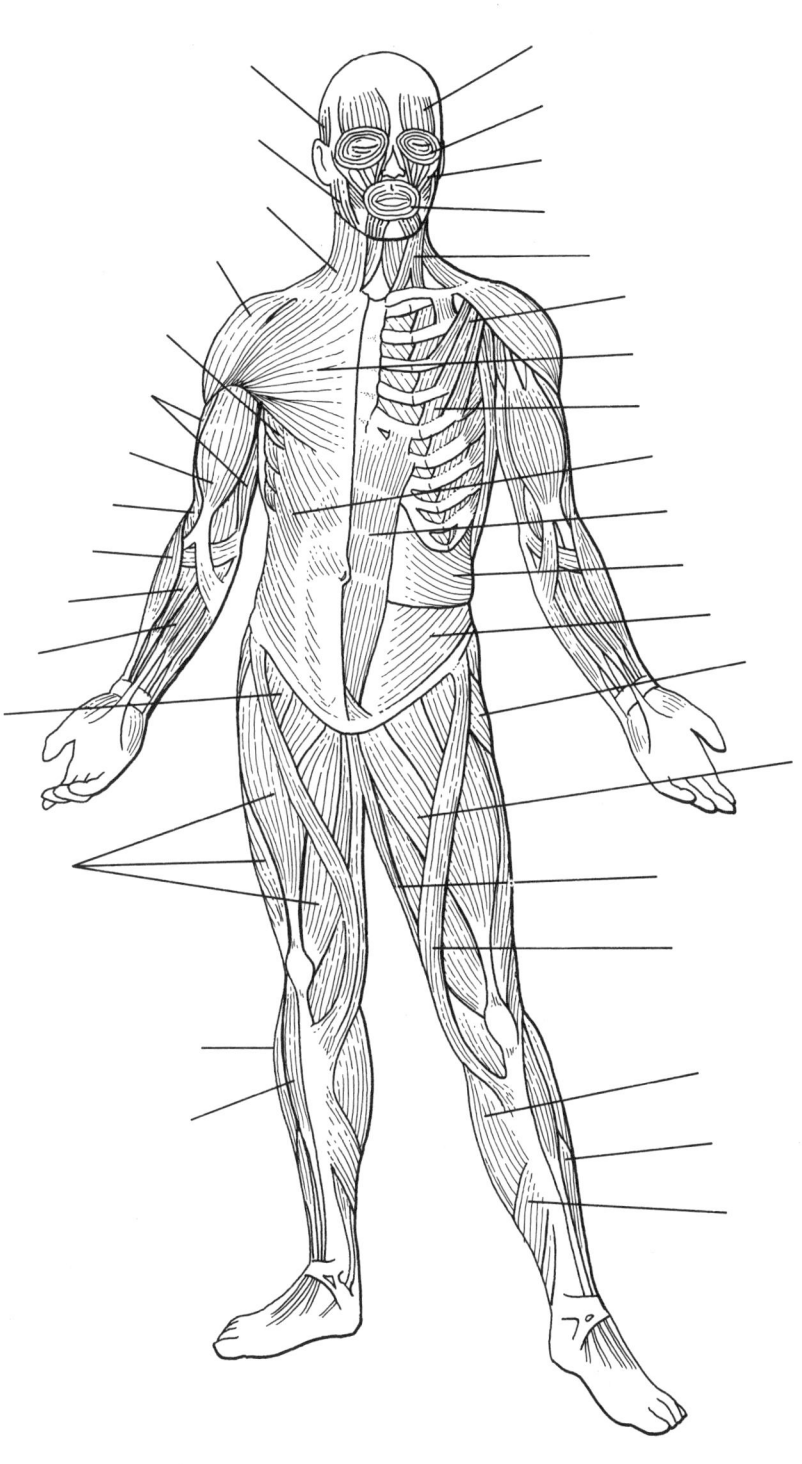

8. Label the muscles. What is the function of each muscle?

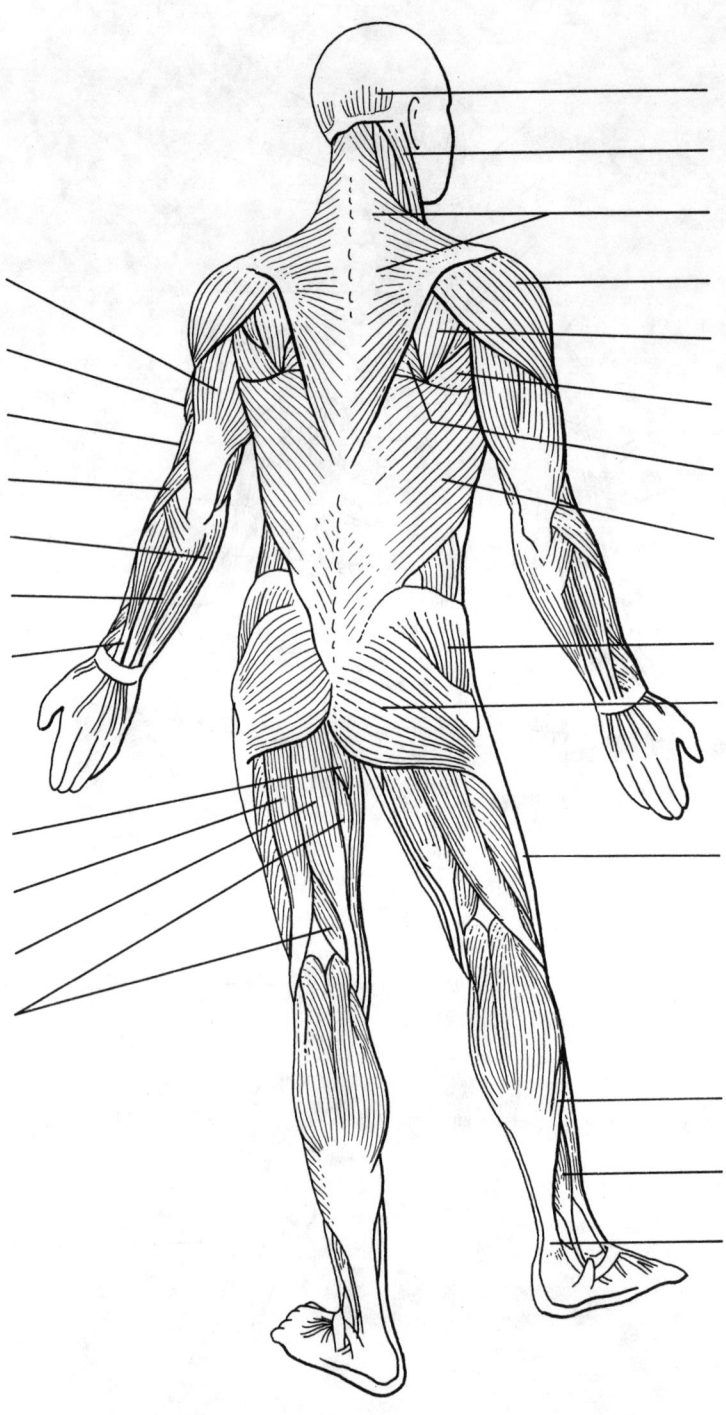

9. Label the muscles. What is the function of each muscle?

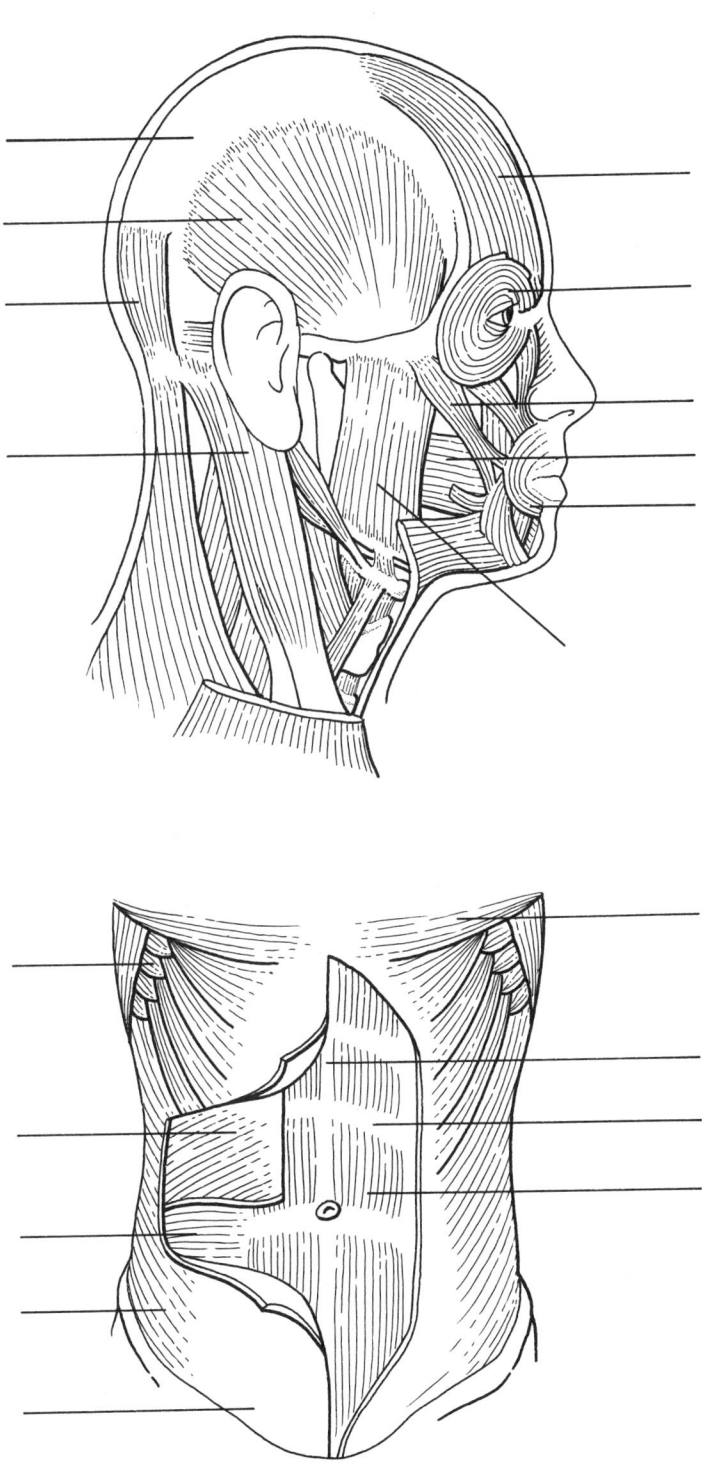

10. Label the muscles. What is the function of each muscle?

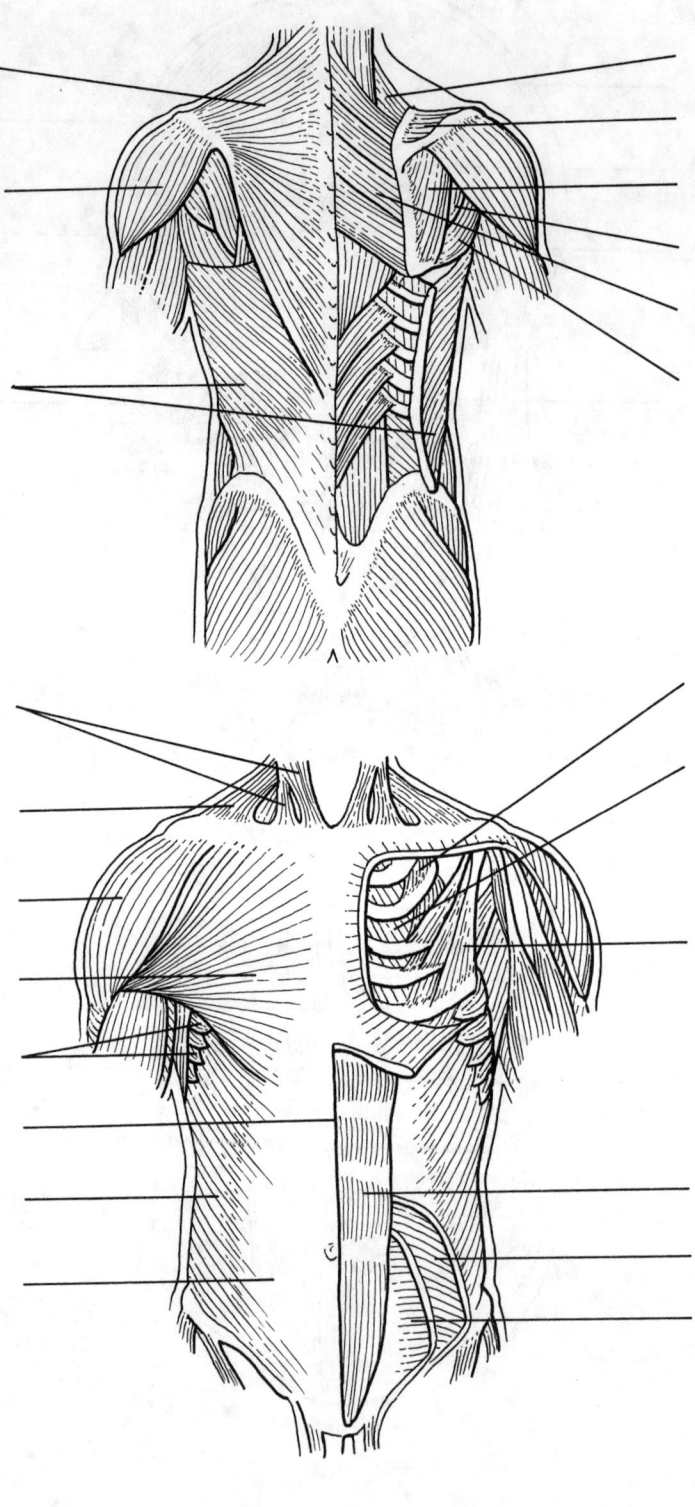

11. Label the muscles. What is the function of each muscle?

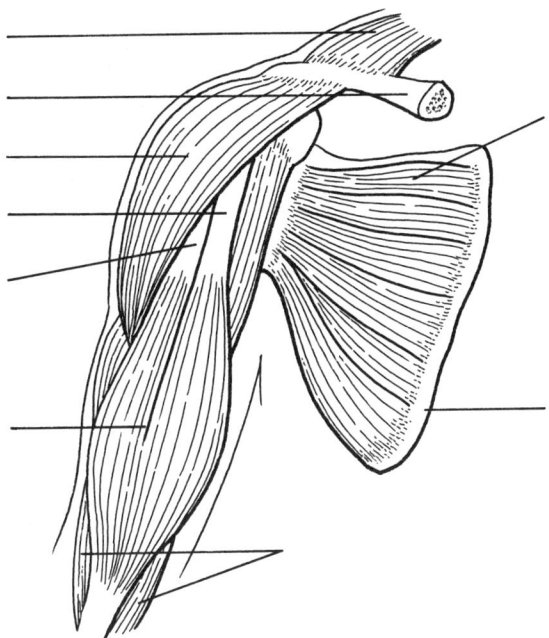

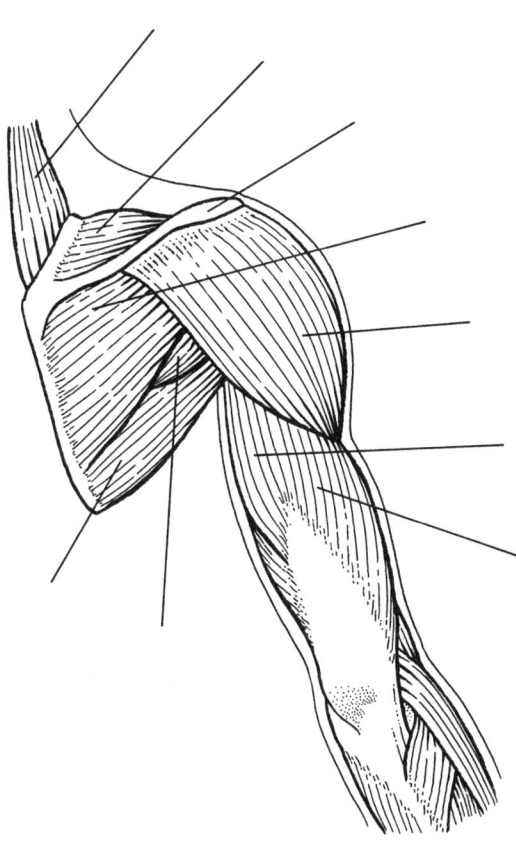

12. Label the muscles. What is the function of each muscle?

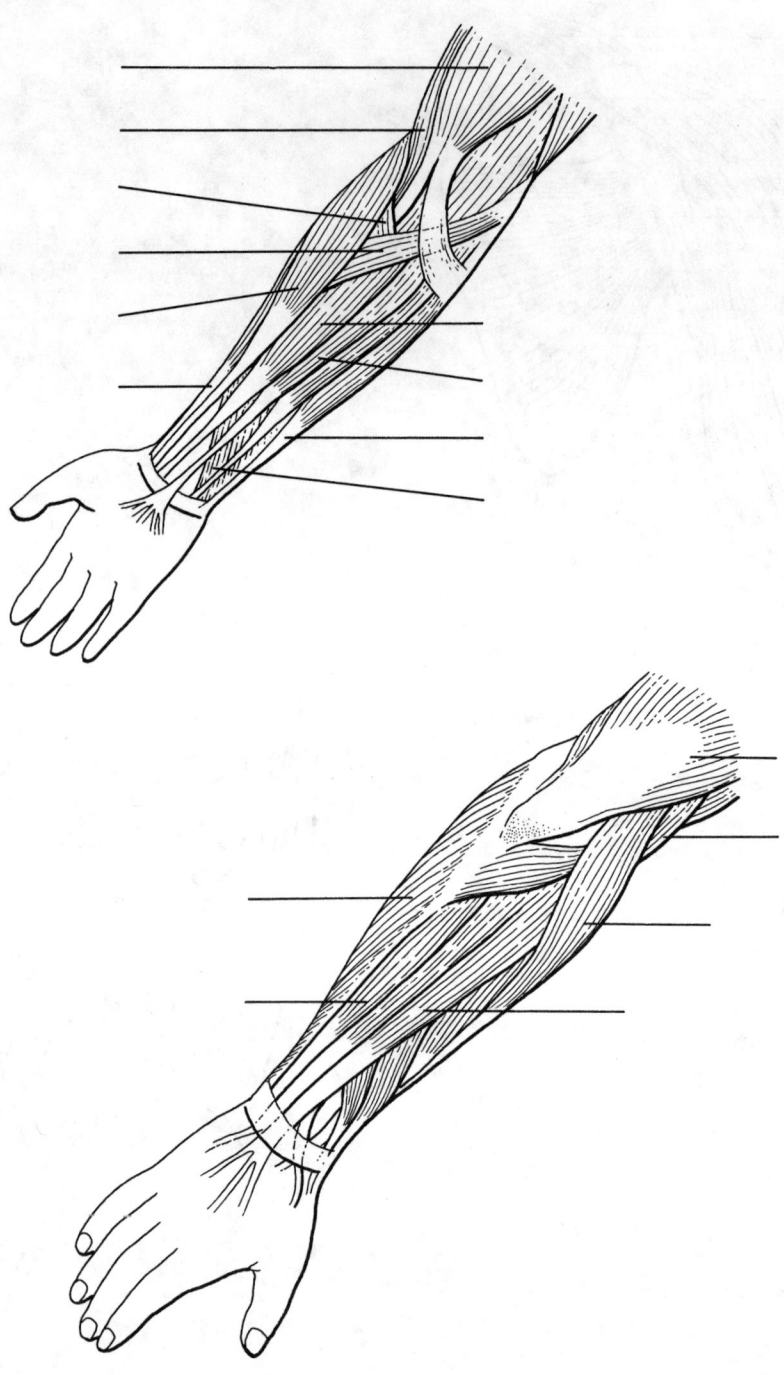

144

13. Label the muscles. What is the function of each muscle?

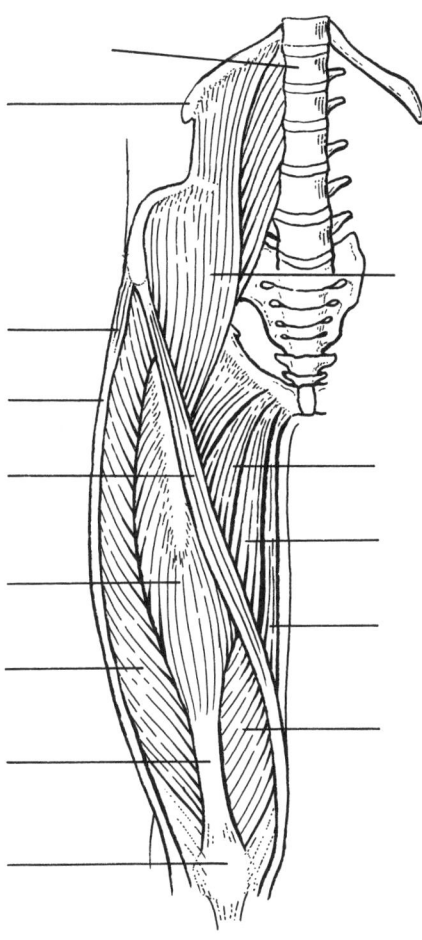

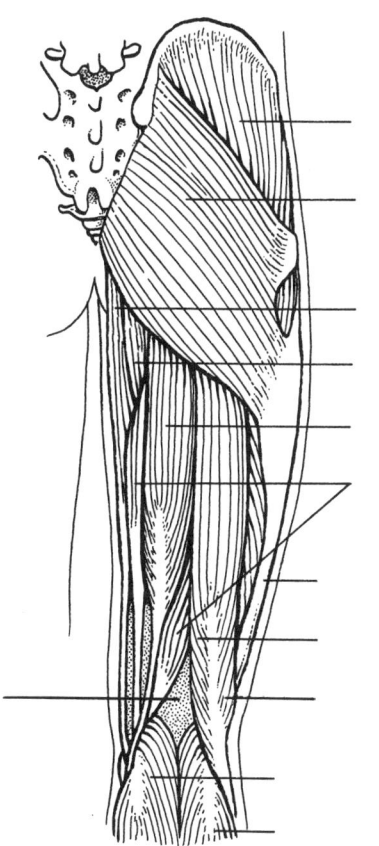

14. Label the muscles. What is the function of each muscle?

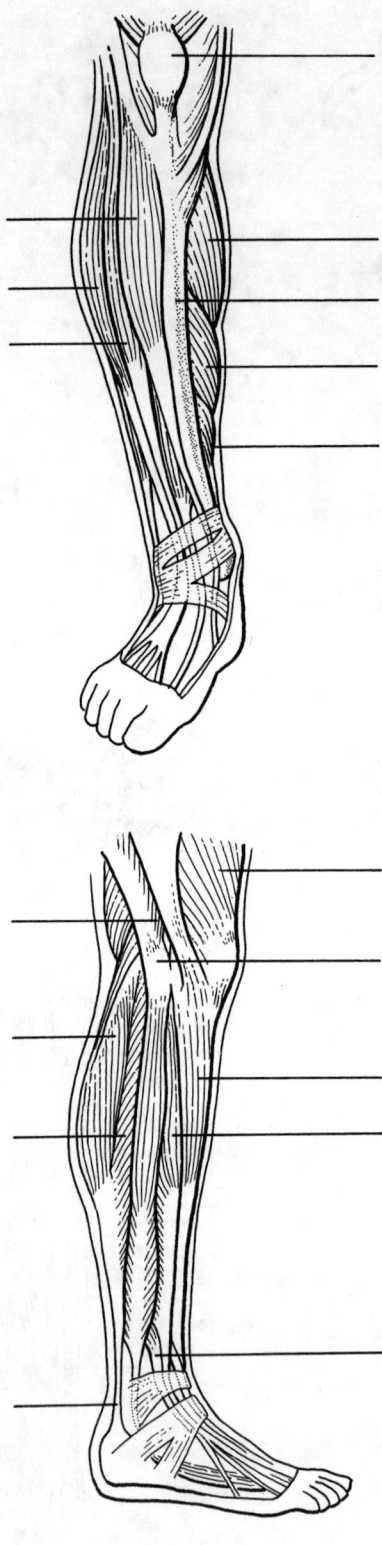

146

QUESTIONS TO MAKE YOU THINK

1. "The most abundant connective tissue associated with muscle is **[deep] fascia** ..." What kind of tissue is found in deep fascia? What features make this tissue effective as an external wrapper for muscles and as an anchor for tendons? (See chapter four.)

2. "A neurotransmitter is a chemical that carries a signal from one nerve terminal to another neuron or a muscle cell." What process that you studied in Chapter three allows the neurotransmitter to bridge the synaptic cleft?

3. "[The sarcoplasmic reticulum] functions in the storage of calcium, which is needed for muscle contraction." The calcium is stored in the form of ions. What is an ion?

    Are calcium ions cations or anions?

    When a muscle fiber is at rest, calcium ions are at higher concentration within the sarcoplasmic reticulum than outside of it. What cellular process allows the cell to concentrate calcium ions against a concentration gradient?

4. "A muscle does not partially contract; it either does it all the way, or not at all. This feature of muscle fibers is called the **all-or-none response**." How can the all-or-none response be reconciled with the obvious ability of a person to alter the intensity of their muscular contractions?

5. Severe Bell's palsy is a disease condition resulting from damage to the facial nerve. One result is the inability to close the eye on the affected side of the body. How can you account for this symptom?

ADDITIONAL STUDY

Read the Chapter Summary (pages 232 -233. Write down the definitions of all the KEY TERMS (page234).

Review the illustrations in your textbook and understand the answers to the questions associated with each one. The answers are on pages 235-236.

After you have studied the chapter, test yourself by **writing** the answers to the "CONCEPTS CHECKS" and "QUESTIONS FOR REVIEW" in your text (pages 203, 208, 212, 214, 229, and 234). Writing the answers forces you to challenge yourself.

The day before an exam over this chapter, read the "Learning Objectives", page 196, and review any of the sections which you think might cause a problem.

CHAPTER 9: ORGANIZATION OF THE NERVOUS SYSTEM

CONTENT MASTERY

DIVISIONS OF THE NERVOUS SYSTEM.

_____ 1. The two major categories of the nervous system include the ... and the peripheral nervous system.

_____ 2. The CNS includes the ... and spinal cord.

_____ 3. All (sensory/motor) impulses arrive into the CNS.

_____ 4. Motor impulses (originate in/arrive into) the CNS.

_____ 5. Nerves that carry impulses to and away from the central nervous system comprise the ...

_____ 6. The ... nerves of the PNS originate from the brain and extend to the head and neck.

_____ 7. The ... nerves of the PNS arise from the spinal cord to supply the body below the head.

_____ 8. The ... subdivision of the PNS includes nerves that are under voluntary control.

_____ 9. The ... system includes mainly nerves that carry impulses to visceral organs, blood vessels, and glands.

_____ 10. The autonomic nervous system is often referred to as the (involuntary/voluntary) nervous system.

_____ 11. The autonomic system is further subdivided into the ... division and the parasympathetic division.

_____ 12. The (sympathetic/parasympathetic) nerves play an active role during conditions of stress.

_____ 13. The (sympathetic/parasympathetic) nerves function during normal organ functioning.

NERVE TISSUE. As you read about nerve tissue, speak aloud everything that you think could be asked as a test question. Hearing yourself say the terms and their definitions helps your memory by allowing input into your brain by way of both your eyes and your ears. After you have read the section, work on the following matching series. This time, there will be several questions with multiple answers, and some terms may not be used at all! Have fun!

astrocytes
axon
cell body
collaterals
dendrites
ependymal cells
gray matter
microglia
myelin sheath
myelinated fibers
neuroglia

neurilemma
neurons
nerve fiber
neurofibrils
nissl bodies
nodes of Ranvier
oligodendrocytes
Schwann cells
unmyelinated fibers
white matter

_____ 14. the primary structural and functional unit of nerve tissue

_____ 15. nervous tissue cell types: neurons and ...

_____ 16. a type of nerve tissue which conducts impulses at great speeds

_____ 17. a type of nerve tissue which supports and maintains neuronal structure and function

_____ 18. type of nerve tissue which forms about 90% of the substance of the brain and spinal cord

_____ 19. types of neuroglia located in the brain and spinal cord

_____ 20. type of neuroglia found in the large nerves of the peripheral nervous system

_____ 21. type of neuroglia which anchors neurons to blood capillaries

_____ 22. type of neuroglia which help form cerebrospinal fluid

astrocyte
axon
cell body
collaterals
dendrites
ependymal cells
gray matter
microglia
myelin sheath
myelinated fibers
neuroglia

neurilemma
neurons
nerve fiber
neurofibrils
nissl bodies
nodes of Ranvier
oligodendrocytes
Schwann cells
unmyelinated fibers
white matter

_____ 23. type of neuroglia which destroy invading microorganisms

_____ 24. type of neuroglia which form myelin sheath in CNS

_____ 25. type of neuroglia which provide insulating coverings around large PNS axons

_____ 26. portion of a neuron consisting of cytoplasm, plasma membrane, nucleus and nucleolus

_____ 27. membranous sacs found in neurons, similar to RER

_____ 28. network of fine threads in neurons

_____ 29. tree-like, thin, branching extensions of the cell body of neurons

_____ 30. fiber conducting impulses away from the cell body of neurons

_____ 31. extensions conducting impulses to the cell body of neurons

_____ 32. extension from the cell body of neurons that may be one meter long

_____ 33. white, fatty insulating barrier found on neurons

_____ 34. literally means "tree"

_____ 35. literally means "nerve sheath"

astrocyte
axon
cell body
collaterals
dendrites
ependymal cells
gray matter
microglia
myelin sheath
myelinated fibers
neuroglia

neurilemma
neurons
nerve fiber
neurofibrils
nissl bodies
nodes of Ranvier
oligodendrocytes
Schwann cells
unmyelinated fibers
white matter

_____ 36. named for the French physician, Louis Ranvier

_____ 37. gaps between Schwann cells where myelin sheath is absent

_____ 38. axons that lack myelin sheath

_____ 39. groups of axons of the brain with white myelin sheaths

_____ 40. groups of unmyelinated fibers and neuron cell bodies in the brain

TYPES OF NEURONS. Select the word or words in parentheses which best completes each of the following. Extra work means extra pay in learning!

_____ 41. The neurons that carry impulses from the CNS to hand muscles to allow you to write your answer are (bipolar, multipolar, unipolar) neurons.

_____ 42. The neurons found in the eyes which allow you to read this question are (bipolar, multipolar, unipolar) neurons.

_____ 43. The neurons that carry heat impulses from skin receptors to the spinal cord are (bipolar, multipolar, unipolar) neurons.

_____ 44. Another term for "sensory neuron" is (afferent neuron, efferent neuron, interneuron).

_____ 45. Association neurons are located in the (CNS, PNS).

_____ 46. Another term for "motor neuron" is (afferent neuron, interneuron, efferent neuron).

_____ 47. The type of neurons which carry nerve impulses from the CNS to salivary glands to make you salivate when you see someone bite into a raw lemon are (sensory neurons, association neurons, motor neurons).

NEURON FUNCTION (THE PHYSIOLOGY OF A NERVE IMPULSE). Many "A and P" courses are mostly "A" and very little "P". This section of text is real physiology! Read through the entire section aloud, studying the figures carefully. Imagine that you will have to tutor someone you have been wanting to impress. After you have read with intense attention, answer the following.

_____48. The primary function of a neuron is
   a. the insulation of an impulse
   b. the generation of a nerve impulse
   c. the production of ATP to provide energy for impulse speed
   d. the formation of cerebrospinal fluid

_____49. Membrane potential is best described as
   a. the use of energy to transport sodium ions out of the cell.
   b. the ability of neurons to respond to a change in their environment.
   c. the use of energy to transport potassium ions out of the cell.
   d. the difference in charge on either side of a cell membrane.

_____50. Integral proteins in the membrane of neurons which form guarded are
   a. membrane potentials
   b. ion channels
   c. cell pores
   d. the sodium-potassium pump

_____51. Ion channels are peculiar to neurons
   a. true
   b. false

_____52. Which ions produce the greatest influence in creating a membrane potential?
   a. sodium and potassium
   b. chloride and potassium
   c. sodium and calcium

_____53. When a neuron is at rest, where is the greatest concentration of sodium ions?
   a. outside the cell
   b. inside the cell
   c. in the ion channels of the membrane

_____54. Consider the sodium-potassium pump. As a result of the combined effects of sodium being pumped out and potassium leaking out, what kind of charge does the outside of the plasma membrane tend to accumulate?
   a. positive
   b. negative
   c. neutral

_____55. The inside of the cell has a negative charge due to which of the following?
   a. low amounts of positive ions
   b. the presence of chloride ions (Cl⁻)
   c. the presence of negatively-charged proteins too large to cross the membrane
   d. all of the above

_____56. An uneven distribution of ions produces a ... in a resting neuron.
   a. negative potential
   b. positive potential
   c. membrane potential
   d. resting potential

_____57. A membrane that exhibits the resting potential is said to be in a polarized state.
   a. true
   b. false

_____58. The depolarization of a region of plasma membrane followed quickly by its repolarization is called
   a. action potential
   b. resting potential
   c. membrane potential
   d. a nerve impulse potential

_____59. Many types of cells exhibit action potentials.
   a. true
   b. false

_____60. Which is an illustration of saltatory conduction
   a. electricity through an insulated wire
   b. electron flow between poles of a battery
   c. a pump
   d. a flat skipping stone across water

_____ 61. Which of the following is least related to the saltatory conduction?
   a. myelinated fibers
   b. unmyelinated fibers
   c. nodes of Ranvier
   d. split-second responses in emergency situations

_____ 62. If a stimulus is strong enough to cause an action potential, the impulse will be conducted along the entire length of the neuron at a maximum strength. This statement describes
   a. action potential
   b. resting potential
   c. saltatory conduction
   d. all or none response

_____ 63. The minimal strength of a stimulus required to initiate an action potential is referred to as the
   a. all or none response
   b. threshold stimulus
   c. subthreshold stimulus
   d. summation

_____ 64. A stimulus weaker than threshold is called
   a. all or none response
   b. threshold stimulus
   c. subthreshold stimulus
   d. summation

_____ 65. A series of weak stimuli (less than threshold) quickly applied may have an effect that can lead to an action potential. This phenomenon is called
   a. all or none response
   b. threshold stimulus
   c. subthreshold stimulus
   d. summation

_____ 66. A synapse
   a. is a cumulative effect that can lead to an action potential
   b. receives an impulse from a presynaptic neuron and passes it to the postsynaptic neuron.
   c. is a junction between adjacent neurons.
   d. Both b and c are correct.

67. Sequence the following events which are associated with the transmission of an impulse across a synapse.
    a. synaptic end bulb of a presynaptic neuron
    b. neurotransmitters contact the cell membrane of postsynaptic neuron
    c. neurotransmitter is inactivated by enzymes
    d. release of neurotransmitters into the synapse
    e. calcium ions flow into the end bulb
    f. postsynaptic neuron responds

    _____ _____ _____ _____ _____ _____

_____ 68. If acetylcholine increases the postsynaptic neuron's permeability to sodium ions and an action potential occurs, this is called
    a. an excitatory transmission
    b. an inhibitory transmission

_____ 69. Norepinephrine is a neurotransmitter which is
    a. excitatory
    b. inhibitory

_____ 70. The speed at which a nerve impulse is conducted varies according to the presence or absence of
    a. axons
    b. dendrites
    c. myelin sheaths
    d. repolarization

_____ 71. Endorphins
    a. inhibit impulses generated by pain stimuli
    b. serve as the body's natural pain killers
    c. are neurotransmitters which cause inhibitory transmission
    d. all of the above

_____ 72. Which of the following are inhibitory neurotransmitters? (You may choose more than one answer.)
    a. norepinephrine
    b. acetylcholine
    c. GABA
    d. enkephalins
    e. endorphins

_____ 73. What would happen if the postsynaptic membrane received acetylcholine from thousands of presynaptic neurons and GABA from hundreds of presynaptic neurons?
a. the accumulated effect would be excitatory, and an action potential would result.
b. the accumulated effect would be inhibitory, and the action potential would be blocked.
c. the cell would respond with a partial action potential
d. the accumulated effect would be confusion, and the postsynaptic membrane would not respond (the all or nothing principle).

CHEMICALS AND THEIR INFLUENCES ON NERVOUS FUNCTION. Match the category of chemicals to their descriptions.

_____ 74. stimulants

_____ 75. depressants

_____ 76. antidepressants

_____ 77. psychedelic drugs

_____ 78. analgesics

a. interfere with the transmission of pain impulses to the brain

b. alter or distort perception by affecting the neurotransmitter serotonin

c. increase synaptic transmission of impulses

d. inhibit impulses at the synapse

e. increase norepinephrine in the brain

THE CENTRAL NERVOUS SYSTEM - THE SPINAL CORD. Match the protective coverings of the spinal cord with their descriptions. Pay attention to spelling as you write the terms in the blanks.

arachnoid
cerebrospinal fluid
dura mater
epidural space

meninges
pia mater
spinal tap
subarachnoid space

_____ 79. literally, "membranes"

_____ 80. outermost membrane around the spinal cord

_____ 81. between dura mater and vertebral column

_____ 82. the middle membrane around the spinal cord which looks like a cobweb

_____ 83. thin, innermost membrane attached to surface of the spinal cord

_____ 84. space between arachnoid and pia mater

_____ 85. clear, colorless fluid which fills the subarachnoid space

_____ 86. literally, "delicate mother"

_____ 87. literally, "tough mama"

_____ 88. a lumbar puncture below the second lumbar vertebra

SPINAL CORD STRUCTURE. Be sure to study the illustrations in your text and review them in the "LABEL AND LIST" section of this study guide. Additional pertinent information concerning the structure are covered here. Answer the following in complete sentences.

89. Considering the literal meaning of conus medullaris, how is its name related to its structure?

90. Considering the literal meaning of the cauda equina, how is its name related to its structure?

91. What can be found in the central canal?

92. There is a functional distinction between the three types of gray horns of the spinal cord. Contrast them in this chart.

SPINAL CORD HORNS	FUNCTION
Anterior gray horn	
Posterior gray horns	
Lateral gray horns	

93. The white matter in the spinal cord is composed mostly of _____ fibers.

SPINAL CORD FUNCTIONS. Match the terms and descriptions associated with the functions of the spinal cord.

ascending tracts
association neurons
descending tracts
motor neurons

reflex arc
sensory neuron
somatic reflexes
visceral reflexes

_____ 94. a nerve tract carrying sensory information to the brain

_____ 95. a nerve tract carrying motor information away from the brain

_____ 96. a simple pathway in which an impulse travels from stimulus, to spinal cord, to motor neuron without traveling to the brain

_____ 97. a neuron receiving sensory stimulation and carrying it to the spinal cord

_____ 98. a neuron that processes information from the sensory, routing it to appropriate motor neurons

_____ 99. a reflex arc affecting skeletal muscles

_____ 100. withdrawal and patellar reflexes, for example

_____ 101. reflexes which affect heart rate, breathing, and sneezing

THE BRAIN. It is difficult to imagine the myriad of functions performed by the brain: the management of breathing, digestion, and heart action, the storage of memories, the integration of information and consciousness, decision making, analysis, etc. With a sense of wonderment and appreciation, focus your brain on itself. It's mind-boggling!

Place each the following major parts of the brain on the chart to indicate their location.

cerebrum   cerebellum   diencephalon   medulla oblongata   pons

Forebrain	Midbrain	Hindbrain
102.		104.
103.		105.
		106.

Read the section "CEREBROSPINAL FLUID AND VENTRICLES OF THE BRAIN." Match the terms with their descriptions.

A. arachnoid villi
B. cerebral aqueduct
C. cerebrospinal fluid (CSF)
D. choroid plexus
E. fourth ventricle
F. foramen of Monro
G. lateral ventricles
H. third ventricle
I. ventricle

_____107. a clear liquid cushion for the brain

_____108. cavities filled with CSF located in each cerebral hemisphere

_____109. CSF-filled cavity in the center of the diencephalon

_____110. CSF-filled cavity between the cerebellum and the medulla

_____111. a channel connecting lateral ventricles to the third ventricle

_____112. connection between the third and fourth ventricle

_____113. site of CSF formation

_____114. reabsorbs CSF back into the bloodstream

_____115. literally, little chamber

Answer the following multiple choice questions pertaining to the same material, "CEREBROSPINAL FLUID AND VENTRICLES OF THE BRAIN."

_____116. Which of the following is NOT a function of the cerebrospinal fluid?
   a. carries blood to neurons of the brain
   b. prevents the brain from smashing against the inside of the skull during sudden body movements
   c. helps to nourish the brain
   d. removes metabolic wastes from the brain

_____117. Where is the cerebrospinal fluid formed?
   a. in the blood plasma
   b. in the ventricles
   c. in the cerebrum
   d. in the lymphatic system

_____118. There is a discriminating function of the brain which permits certain substances (such as food, water, and oxygen) to enter it while prohibiting the entrance of such things as bacteria. This is called
   a. the phenomenon of Monro
   b. CSF prohibition
   c. choroid plexus
   d. the blood-brain barrier

_____119. CSF is produced continuously.
   a. true
   b. false

_____120. CSF flows in one direction.
   a. true
   b. false

121. Sequence the path of CSF from production to reabsorption into the bloodstream by rearranging the following:
   a. cranial subarachnoid space
   b. through the foramen of Monro
   c. in the lateral ventricles
   d. into the bloodstream
   e. in the third ventricle
   f. the central canal and subarachnoid space of the spinal cord
   g. into the fourth ventricle
   h. through the cerebral aqueduct

_____ _____ _____ _____ _____ _____ _____ _____

_____122. The two types of fluids within the cranial cavity are
    a. water and cerebrospinal fluid both located in the ventricles
    b. blood and cerebrospinal fluid both located in the ventricles
    c. blood in blood vessels and cerebrospinal fluid in ventricles
    d. blood in vessels and water in the ventricles

_____123. Which is NOT true of hydrocephalus?
    a. It reduces cranial pressure.
    b. It is called "water on the brain".
    c. Treatment involves draining excess CSF.
    d. It is seen in infants and is a congenital defect.

THE STRUCTURES OF THE BRAIN. Scan the sections of text from "CEREBRUM" through "CEREBELLUM". Match the structure with their descriptions.

cerebellum   **medulla oblongata**
cerebrum   **midbrain**
diencephalon   **pons**

_____ 124. largest structure in the brain

_____ 125. posterior portion of the hindbrain

_____ 126. composed of thalamus and hypothalamus

_____ 127. part of the brain stem continuous with the spinal cord

_____ 128. bridges the cerebrum and cerebellum

_____ 129. between diencephalon and pons

CEREBRUM. Fill in the blanks with the appropriate term from your text.

130. The cerebrum is often called the "_____ _____" because of its complex functions in conscious thought, memory, and learning.

131. The cerebrum is divided into right and left _____ _____.

132. The cerebrum has wrinkles known as _____.

133. Upward foldings of the cerebrum are called _____.

134. Shallow grooves in the cerebrum are called _____.

135. The deep groove dividing the right and left cerebral hemispheres is the _____ _____.

136. A deep groove which divides the lower margin of the cerebrum from the cerebellum is the _____ _____.

137. The cerebrum is divided into functional regions called _____.

138. The external region (cortex) of the cerebrum is composed of _____ matter.

139. The main connection between the left and right hemispheres of the cerebrum is the __ _____ _____.

DIENCEPHALON. Have you seen the game in which people are given a time limit to view items on a tray? The person who can list the most items (without looking at the tray again) is the winner. In a similar way, give yourself 5 minutes to read and study the one simple paragraph on the diencephalon, set aside the text, and answer the following.

_____ 140. Correctly spell that portion of the forebrain located inferior to the corpus callosum.

_____ 141. What are the two prominent structures of this part of the brain?

_____ 142. Is the diencephalon made mostly of gray or white matter? (choose one)

_____ 143. Diencephalon literally means ....

THALAMUS AND HYPOTHALAMUS. Match the following.

hypothalamus   pineal gland   pituitary gland   thalamus

_____ 144. the largest portion of the diencephalon

_____ 145. small oval structure associated with the thalamus

_____ 146. controls body temperature

_____ 147. principle relay station for incoming sensory impulses

_____ 148. principal relay station for outgoing, involuntary motor impulses

hypothalamus **pineal gland** **pituitary gland** **thalamus**

_____ 149. associated with aggression

_____ 150. partially housed in bone

_____ 151. associated with the pituitary gland

_____ 152. regulates food intake

_____ 153. controls wake-sleep patterns

_____ 154. signals hunger and thirst

MIDBRAIN AND PONS. Match the following.

cerebral peduncles **colliculi** **midbrain** **pons**

_____ 155. contains colliculi

_____ 156. bundles of myelinated fibers in the midbrain

_____ 157. reflex centers in the midbrain for rapid eye movements

_____ 158. literally, a bridge

_____ 159. made mostly of white matter with scattered masses of nuclei

MEDULLA OBLONGATA AND CEREBELLUM.

medulla oblongata **vermis** **sulci**
cerebellum **arbor vitae** **transverse fissure**
brain stem **folia** **cerebellar peduncles**
pyramids

_____ 160. literally, little, main part of the brain

_____ 161. literally, long middle

_____ 162. an up-folding convolution of the cerebellum

medulla oblongata vermis sulci
cerebellum arbor vitae transverse fissure
brain stem folia cerebellar peduncles
pyramids

_____ 163. connecting "worm" between the two hemispheres of the cerebellum

_____ 164. pattern of white matter in cerebellum

_____ 165. a down-folding convolution of the cerebellum

_____ 166. three paired bundles of myelinated fibers in the cerebellum

_____ 167. acts as an "automatic pilot" for motor responses

_____ 168. midbrain, pons, and medulla oblongata, collectively

_____ 169. contains a cardiac center to regulate heart rate

_____ 170. contains the vasomotor center

THE PERIPHERAL NERVOUS SYSTEM. The nerves, ganglia, and sensory receptors which extend outside the CNS are considered the peripheral nervous system. It's easy to remember because "periphery" refers to "the side". Glance through the outline for this section before you begin reading to get the scope of this material.

Match the parts of the peripheral nervous system with their descriptions.

autonomic system mixed nerve perineurium
endoneurium motor nerve sensory nerve
epineurium nerve somatic system
ganglia nerve fiber

_____ 171. a subdivision of the PNS which is involved with conscious activities

_____ 172. a subdivision of the PNS involved with unconscious activities

_____ 173. literally, pertaining to the body

autonomic system   mixed nerve   perineurium
endoneurium   motor nerve   sensory nerve
epineurium   nerve   somatic system
ganglia   nerve fiber

_____ 174. literally, self regulating

_____ 175. portion of the PNS which controls the muscles you are using to write

_____ 176. the axon of a single neuron

_____ 177. parallel bundles of nerve fibers enclosed in connective tissue

_____ 178. literally, upon a nerve

_____ 179. outer covering around a nerve

_____ 180. loose, delicate sheath surrounding each nerve fiber

_____ 181. sheath of connective tissue around fascicles of a nerve

_____ 182. a nerve containing only sensory fibers

_____ 183. afferent nerve

_____ 184. efferent nerve

_____ 185. a nerve which carries impulses only away from the CNS

_____ 186. a nerve containing only motor fibers

_____ 187. nerve containing fibers from both sensory and motor neurons

_____ 188. clusters of neuron cell bodies located outside the CNS

CRANIAL NERVES. Ask your instructor specifically what you are to know concerning the cranial nerves. Use Table 9-2 to learn these nerves and their functions. Cover each column as you review to see if you can recite the answers. As a summary of pertinent information not covered in the table, answer the following.

189. How many pairs of cranial nerves are there?

190. Explain how the roman numerals assigned to the cranial nerves are important to your study.

191. What three pairs of cranial nerves contain only sensory fibers?

192. Which two pairs of cranial nerves are entirely motor in function? (They supply muscles of the head and neck.)

STUDY AID. A mnemonic device which has been used for generations to help in memorization of the cranial nerves in order is:

On Old Olympian Towering Tops, A Finn And German Viewed Some Hops.

Each word in the sentence begins with the same letter as one of the cranial nerves:
On - Olfactory I, Old - Optic II, Olympian - Oculomotor III, Towering - Trochlear IV, Tops - Trigeminal V, A - Abducens, Finn - Facial VII, And - Acoustic VIII (This is the old name for the vestibulocochlear nerve), German - Glossopharyngeal IX, Viewed - Vagus X, Some - (Spinal) Accessory XI, Hops - Hypoglossal XII. Some students find this mnemonic device very useful, others find it more difficult than merely memorizing the nerves themselves. You will have to decide if it is useful for you or not.

SPINAL NERVES. Answer the following using the terms provided. You will find it helpful to use this section of study questions as a "reading pacer". Answer as you read through the material. Though the choices are alphabetized, the questions appear in the order they are discussed in the text.

anterior horns (of the gray matter of the spinal cord)
brachial plexus
cervical plexus
communicating rami
dorsal ramus
dorsal root ganglion
dorsal roots
eight
five
lumbosacral plexus
one
phrenic
plexuses
posterior horns (of the gray matter of the spinal cord)
thirty-one
thousands
twelve
ventral ramus
ventral roots

_____ 193. number of pairs of spinal nerves

_____ 194. number of nerve fibers in each spinal nerve

_____ 195. number of pairs of cervical spinal nerves

_____ 196. number of pairs of thoracic nerves

_____ 197. number of pairs of lumbar nerves

_____ 198. number of pairs of sacral nerves

_____ 199. number of pairs of coccygeal nerves

_____ 200. location of cell bodies of motor neurons

_____ 201. location of terminal end of the axons of sensory neurons

_____ 202. sensory fibers between the spinal cord and the point of spinal nerve formation

_____ 203. motor fibers between the spinal cord and the point of spinal nerve formation

_____ 204. large swelling of the dorsal root of each spinal nerve containing cell bodies

anterior horns of the
spinal cord
brachial plexus
cervical plexus
communicating rami
dorsal ramus
dorsal root ganglion
dorsal roots

eight
five
lumbosacral plexus
one
phrenic
plexuses
posterior horns of the

spinal cord
thirty-one
thousands
twelve
ventral ramus
ventral roots

_____ 205. posterior division of a spinal nerve outside the vertebral column which supplies back muscles and skin

_____ 206. an anterior division of the spinal nerve outside the vertebral column which supplies the trunk and limbs

_____ 207. small branches from the twelve thoracic and first two lumbar spinal nerves

_____ 208. complex networks of nerves from more than one spinal nerve root

_____ 209. nerves arising from the cervical plexus which control breathing

_____ 210. a network of nerves whose major branches supply the skin and muscles of the upper limbs

_____ 211. a network of nerves whose major branches supply the skin and muscles of the abdominal wall, pelvic wall, and lower limbs

SOMATIC SYSTEM. Scratch your ear. That effort was accomplished largely by the somatic nervous system which is responsible for the peripheral nerves under conscious control. After you read the two paragraphs for this section, decide whether each of the statements is true or false. Correct false statements by rewriting the sentence and correcting the italicized word.

212. The *somatic* nervous system controls walking.

213. In the somatic nervous system's sensory pathway, there is *only one* sensory neuron between the sensory receptor and the CNS.

214. The is *only one* type of neurotransmitter released at synapses with skeletal muscles.

215. The motor pathway consists of *one* motor neuron extending between the spinal cord and the effector.

AUTONOMIC SYSTEM. Read about the autonomic nervous system in the outline at the end of this chapter to organize your thoughts. Then carefully read this section of text. Refer back to the outline as often as necessary to keep the material organized in your mind. After reading, answer these brief questions.

216. How was the autonomic system named?

217. What is the function of the autonomic nervous system?

218. Describe the "simple" sensory pathways of the autonomic system.

219. What two basic pathways comprise the autonomic motor pathways?

Carefully read through the material on the autonomic motor pathways and functions. Read through it again. Fill in the following table as you read through the material a third time.

220.	Sympathetic Autonomic System	Parasympathetic Autonomic System	Somatic Motor System (for comparison)
Number of neurons in pathway			
Effector			
Description of preganglionic neuron			
Description of postganglionic neuron			
Location of ganglia			
Neurotransmitter released by postganglionic fiber			

Decide whether each of the following statements is true or false. Correct false statements by rewriting the sentence and correcting the italicized word.

221. Most sympathetic postganglionic fibers release *norepinephrine*.

222. If a fiber releases noradrenalin, it is referred to as an *adrenergic* fiber.

223. Parasympathetic postganglionic fibers release *acetylcholine*.

224. Parasympathetic postganglionic fibers are called *cholinergic* fibers.

225. The *parasympathetic* division of the autonomic nervous system is called the "fight or flight" division.

226. The *parasympathetic* division of the autonomic nervous system is called the "rest-repose" division.

227. *All* visceral organs are innervated by both the sympathetic and parasympathetic autonomic systems.

228. Autonomic regulation *cannot be* achieved by varying the strength of the stimulus.

229. The control centers in the brain for the autonomic nervous center are located in the *cerebrum and cerebellum*.

HOMEOSTASIS. Homeostasis is a fundamental concern for health and a focal point for this course. As you read this material, make up an essay question over this essential topic. Imagine that you have been asked to write a closing page for this chapter in the book. What would you say about homeostasis? Part of learning to write good essay answers is the ability to create good essay questions in advance!

CLINICAL TERMS OF THE NERVOUS SYSTEM. Match the term with its description. The selections may be used more than once.

amyotrophic lateral sclerosis **glioma**
bacterial meningitis **Guillain-Barre syndrome**
cephalalgia **multiple sclerosis**
encephalitis **schizophrenia**

_____ 230. a motor neuron disease, Lou Gehrig's disease

_____ 231. idiopathic polyneuritis

_____ 232. headache

_____ 233. bacterial infection of the meninges

_____ 234. tumors of the nervous system that originate from neuroglial cells

_____ 235. loss of myelin protecting spinal nerves and nerve roots

_____ 236. bacterial infection of brain

_____ 237. abnormal behavior, inability to distinguish reality from dreams

_____ 238. progressive, fatal disease involving deterioration of the myelin sheath protecting axons in the brain.

LABELS AND LISTS

1. List five types of neuroglia.

    a.

    b.

    c.

    d.

    e.

2. One the basis of structural differences, there are three major types of neurons. List them. Sketch each one.

    a.

    b.

    c.

3. What are the three meningeal membranes that protect the spinal cord?

    a.

    b.

    c.

4. List three ways that the spinal cord is protected.

    a.

    b.

    c.

5. The brain is divided into what three major regions?

   a.

   b.

   c.

6. The brain is protected from injury by what three structures?

   a.

   b.

   c.

7. List the three meninges of the brain.

   a.

   b.

   c.

8. List four lobes within each hemisphere of the cerebrum.

   a.

   b.

   c.

   d.

9. Label this drawing of a multipolar neuron.

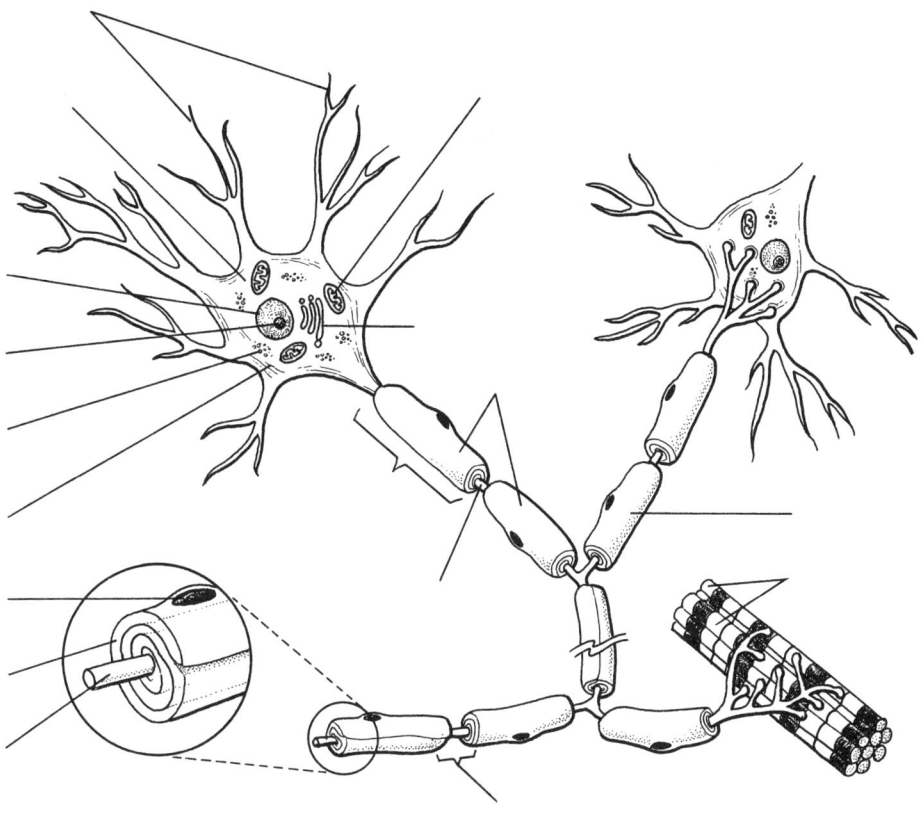

10. Draw a picture of an action potential. Label the resting potential, depolarization, and repolarization. Include the voltage and time measurements. What events are taking place at the cell membrane at each stage?

11. Label the spinal cord and associated structures.

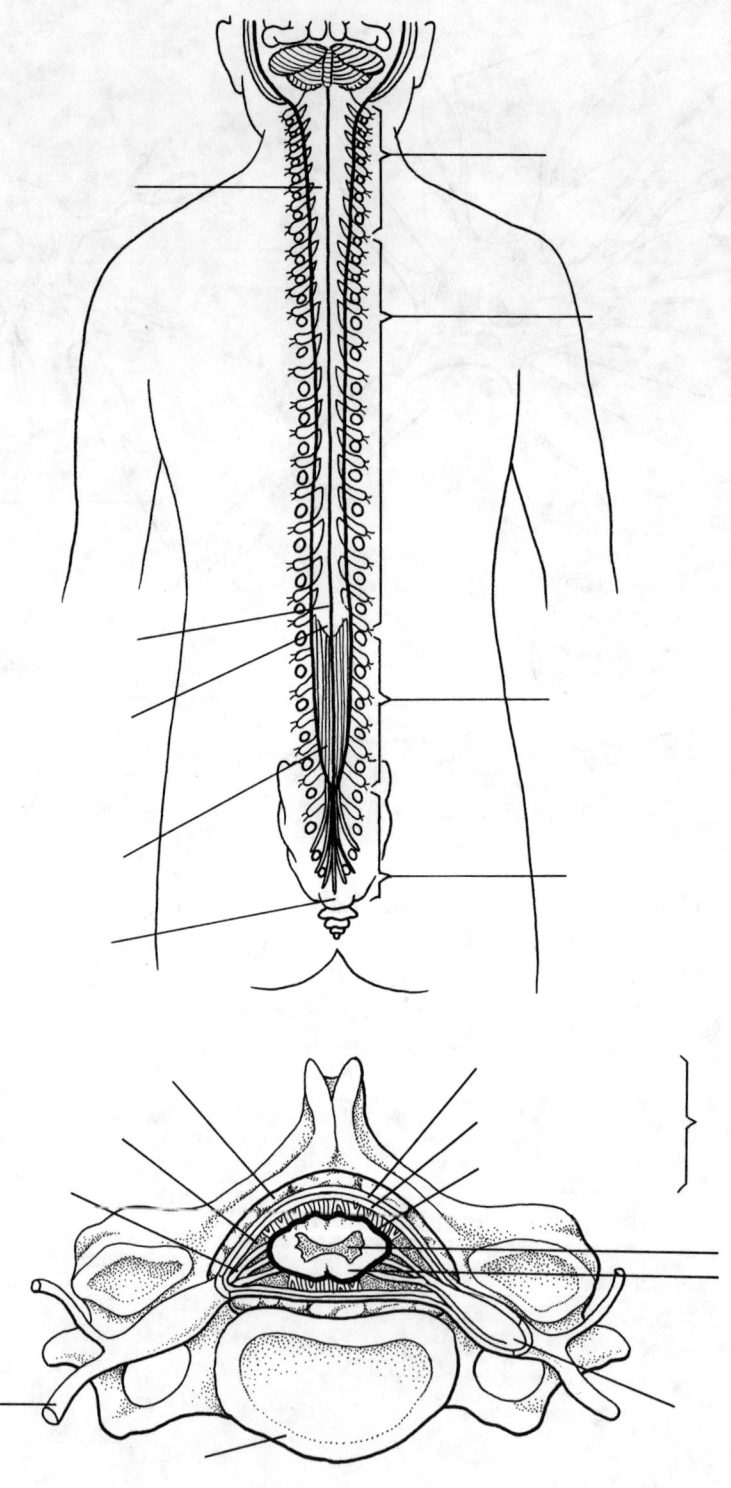

12. Label this drawing of the spinal cord.

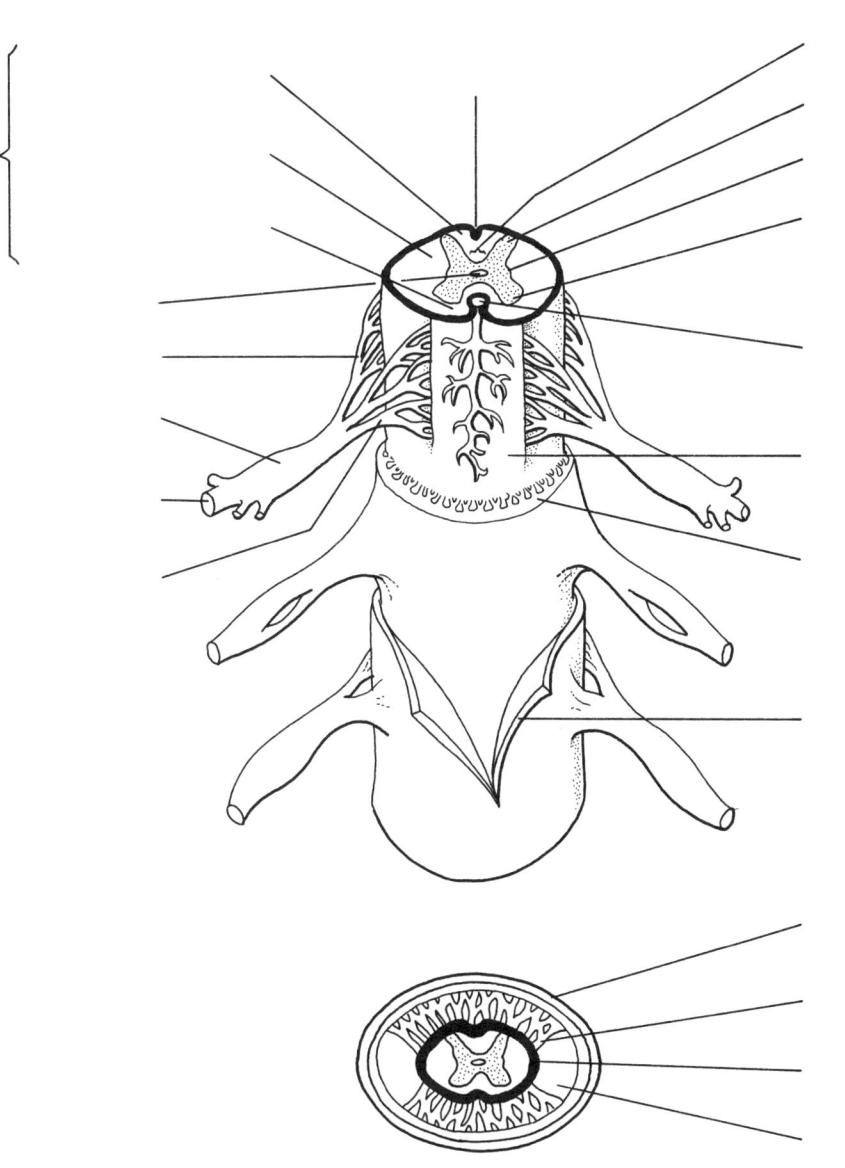

13. Label the brain.

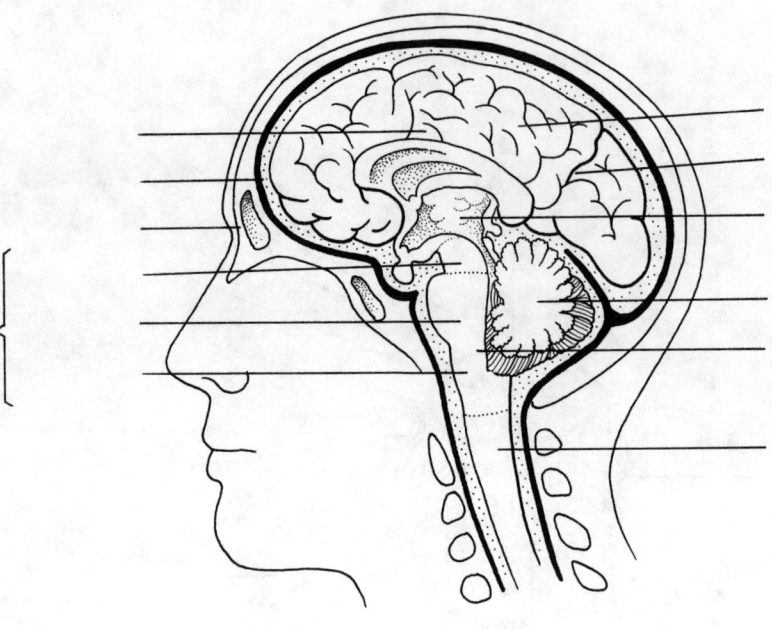

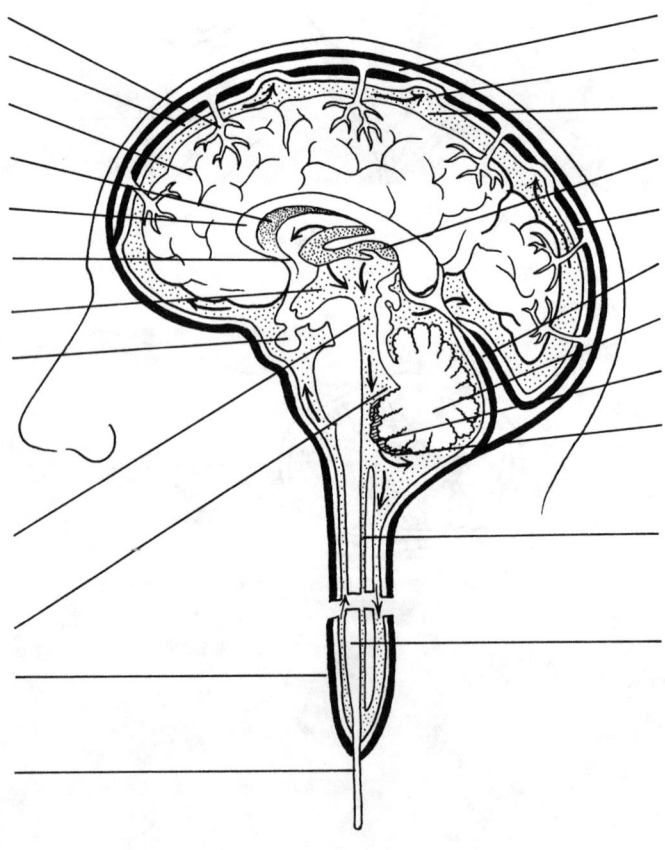

14. Label the brain.

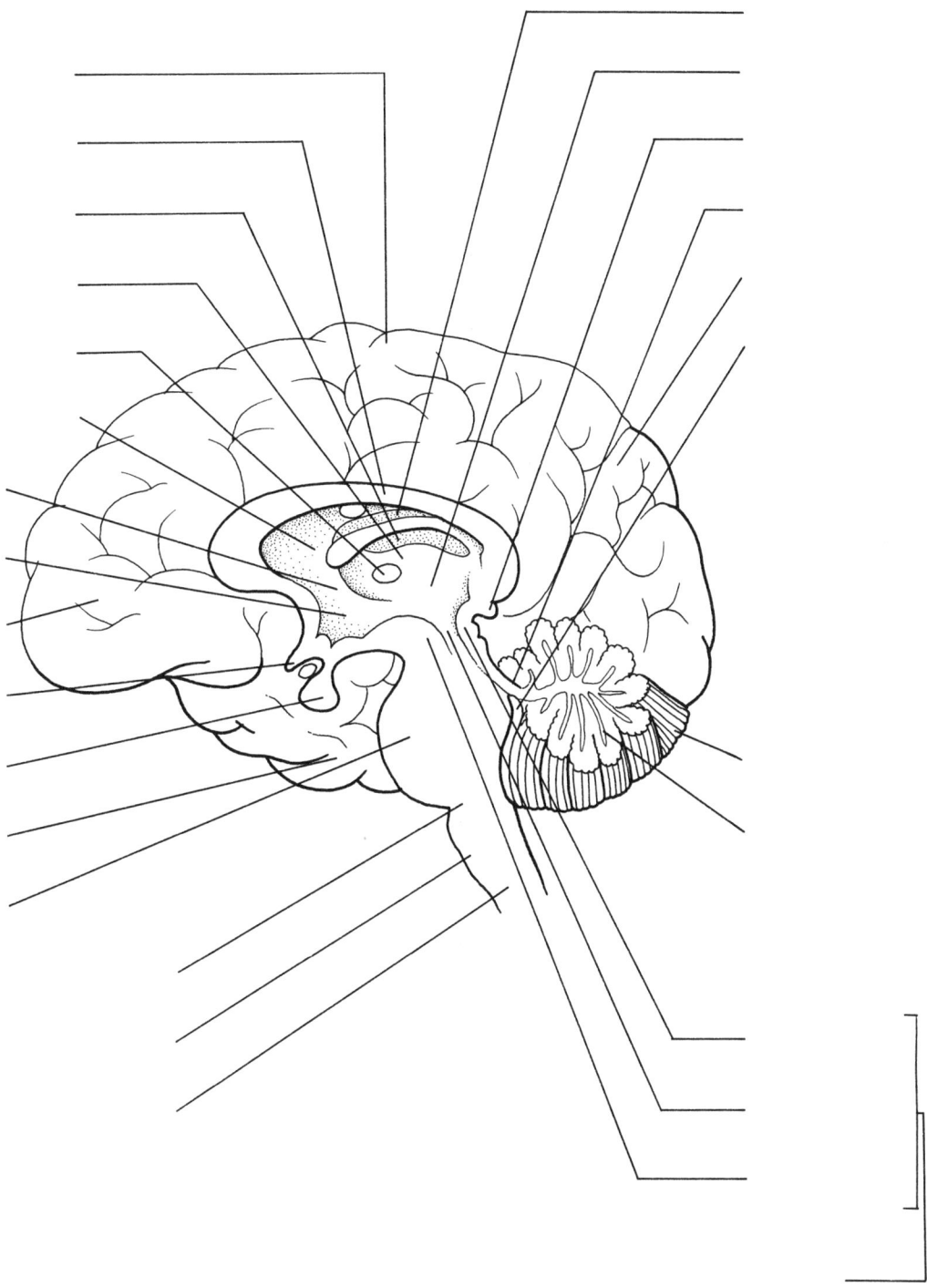

15. Label the nerve.

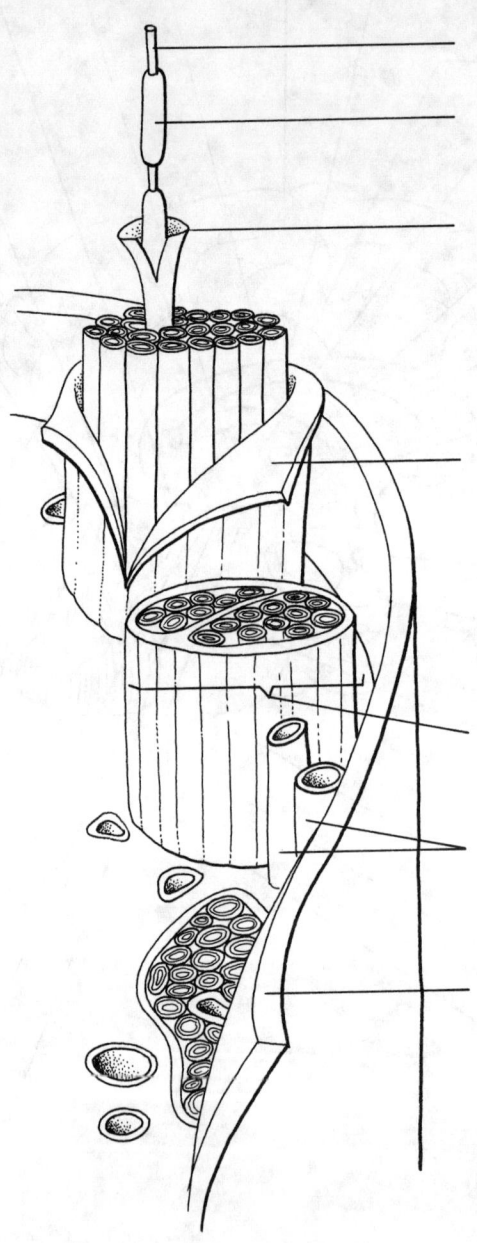

16. Label the cranial nerves, plexuses, and spinal nerves. Label the cranial nerves with their correct names and numbers.

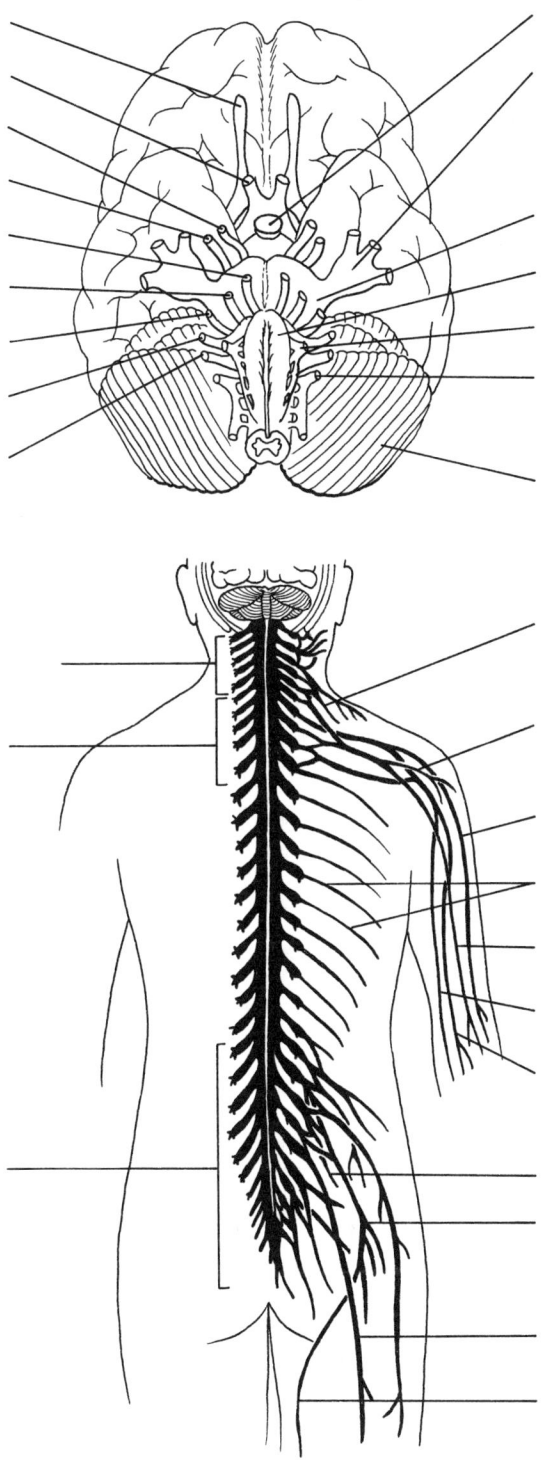

183

QUESTIONS TO MAKE YOU THINK.

1. In Chapter Four you learned about the processes of diffusion and active transport. How do these processes affect a nerve cell?

2. "Like the contraction of muscle fibers, the conduction of a nerve impulse occurs in an **all-or-none** fashion." How is the all-or-none response of muscle cells like that of nerve cells and how is it different?

3. A membrane potential of -70 mV exists across the membrane of a neuron. This potential is established primarily through the action of the Na^+-K^+ pump. Since both of these ions are positively charged, how is it possible for the cell to create a negative potential?

4. "The proteins that form about one-half the composition of the plasma membrane are of two types: **peripheral** and **integral**." (Chapter Three) What kind of proteins are the ion channels and pumps of the nerve cell membrane? Based upon their function, how could you prove this?

ADDITIONAL STUDY

Review the illustrations in your textbook and understand the answers to the questions associated with each one. The answers are on pages 278-279.

After you have studied the chapter, test yourself by **writing** the answers to the "CONCEPTS CHECKS" located throughout the chapter and the "QUESTIONS FOR REVIEW" at the end of the chapter. Writing the answers forces you to challenge yourself.

The day before an exam over this chapter, read the "Learning Objectives", page 238, and review any of the sections which might cause a problem.

NERVOUS SYSTEM
Robert W. Bauman, Jr., Ph.D.

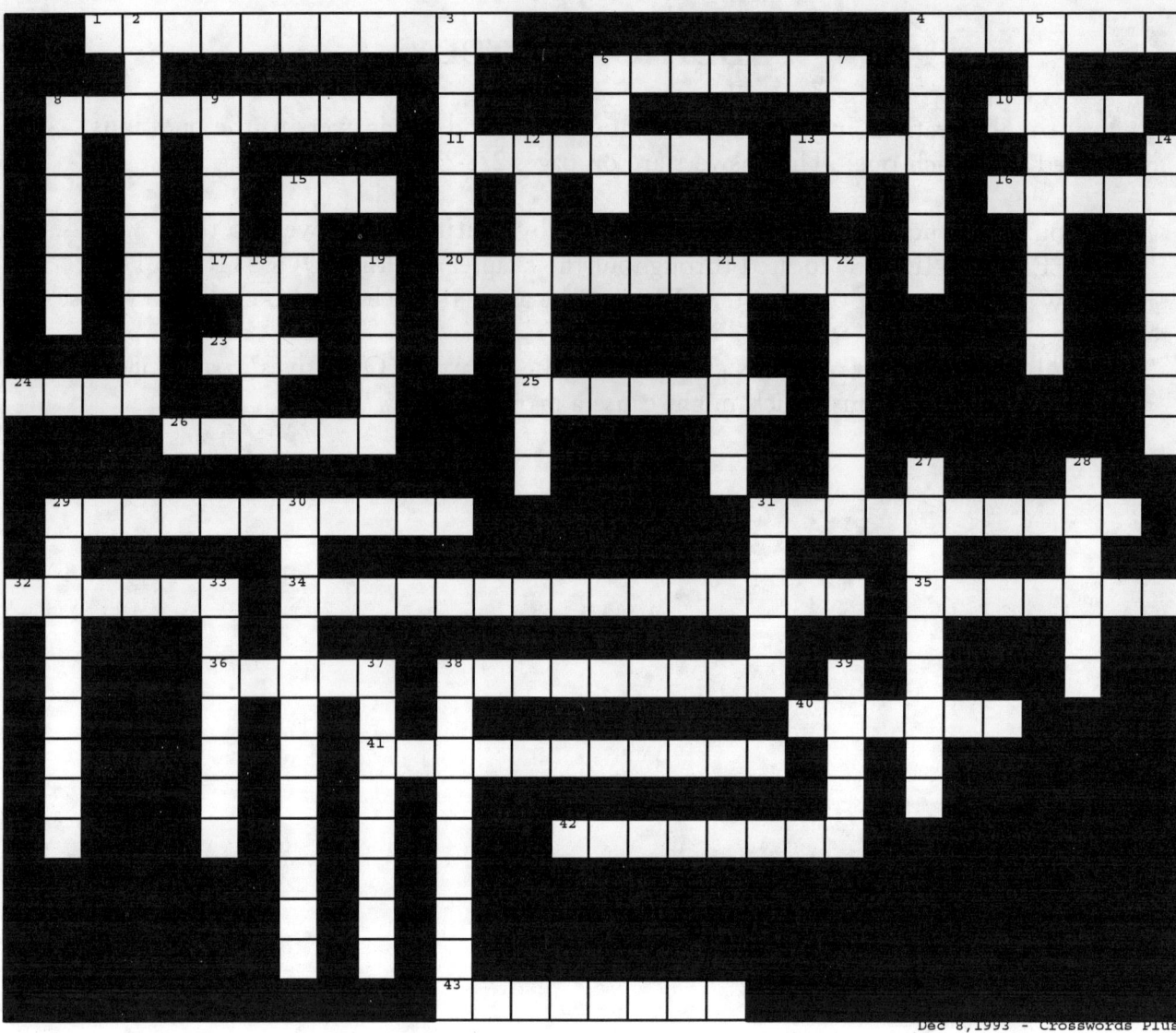

Dec 8, 1993 - Crosswords Plus

Across

1. processing function
4. junction
6. group of cell bodies
8. glial cells
10. spinal nerve branches
11. relay station for most sensory signals
13. bridge
15. delicate membrane
16. myelinated matter
17. brain and spinal cord
20. secrete acetylcholine
23. circulating fluid
24. tough, outer membrane
25. contains pyramids
26. groove
29. autonomic neurons from thorax
31. many dendrites, one axon
32. networks of spinal nerve fibers
34. autonomic division with terminal ganglia
35. gap in myelin sheath; node of ...
36. neuron controlling muscle
38. fourth cranial nerve
40. fatty insulation
41. contains arbor vitae
42. protective membranes
43. conducts impulse toward cell body

Down

2. outer layer of myelin sheath
3. cavity within cerebrum
4. perceptive function
5. middle membrane
6. unmyelinated matter
7. carries signal away from cell body
8. primary functional unit of nerve tissue
9. cranial nerve two
12. controls nonconscious activities
14. largest portion of brain
15. peripheral system
18. similar to RER
19. cranial nerve ten
21. rapid response
22. nerves from brain
27. phagocytic nervous cells
28. cranial nerve seven
29. 'jumping' conducton
30. body's thermostat
31. efferent neuron
33. related to conscious activities
37. generates action potentials
38. minimum stimulus
39. upward fold

CHAPTER 10: THE SPECIAL SENSES AND FUNCTIONAL ASPECTS OF THE NERVOUS SYSTEM

Having mastered the anatomy and basic physiology of the nervous system, move on to the ways in which the nervous system allows you to interact and respond.

CONTENT MASTERY

SENSORY PATHWAYS. Use the list only on the initial work on this section. After you correct your answers, practice filling in the blanks without the aid of the list.

action potential	photoreceptor
chemoreceptor	receptor
general sensory pathway	sensory adaptation
mechanoreceptor	special sensory pathway
nociceptor	stimulus
one	stimulus-specific
	threshold
	three
	thermoreceptor

_____ 1. A change in the environment that is great enough to begin a nerve impulse is called a ...

_____ 2. A stimulus is converted to a nerve impulse by a ...

_____ 3. The minimal stimulus possible to generate an action potential is the ...

_____ 4. Receptors are sensitive to a particular environmental change but insensitive to all others, they are said to be ...

_____ 5. When a sensation disappears as receptors are continuously stimulated, the phenomenon is called...

_____ 6. Receptor that detects mechanical change is a...

_____ 7. Receptor that detects temperature changes is a...

_____ 8. Type of receptor that detects pain from chemical damage to nearby cells is called a ...

_____ 9. Type of receptor that is sensitive to changes in the amount of light is called ...

_____ 10. The type of receptor that detects chemicals dissolved in fluid and which provide smell and taste is a ...

_____ 11. Conduction pathway that carries impulses from a simple receptor such as those in the skin to the brain is known as ....

_____ 12. The number of sensory neurons found in a general sensory pathway to the brain is ...

_____ 13. A conduction pathway that carries an impulse from the eye is called a ...

action potential
chemoreceptor
general sensory pathway
mechanoreceptor
nociceptor
one
photoreceptor
receptor
sensory adaptation
special sensory pathway
stimulus
stimulus-specific
threshold
three
thermoreceptor

GENERAL SENSES. Fill in the blanks with the appropriate term. A list is given, but it is to be used only on the initial work on this section. After you correct your answers, practice filling in the blanks without the aid of these memory boosters.

cerebral cortex
cutaneous
Meissner's corpuscles
Pacinian corpuscles
proprioceptor
referred
visceral

_____ 14. The impulses of the general sensory pathway travel to the ... in the brain where sensation is interpreted and processed.

_____ 15. Touch and pressure are often called ... sensations.

_____ 16. ... are small capsules in the skin which, when moved even slightly, initiate an impulse.

_____ 17. Special receptors which are stimulated by the sense of pressure are called...

_____ 18. Pain that is generated by pain receptors located in the walls of the stomach is known as ... pain.

_____ 19. Pain originating from the heart is felt in the left shoulder, is called ... pain.

_____ 20. A receptor that detects body position is a ...

SPECIAL SENSES - SMELL AND TASTE. Much of the material for this section is covered in the labeling section of this study guide. Additional pertinent information is found here. Read the sections, label the diagrams in the "Label and List" section of this study guide, then select the best choices for each statement here.

_____21. The sense of smell is also known as
   a. olfaction
   b. gustation
   c. vision
   d. lacrimation

_____22. "Sniffing" increases the awareness of smells because
   a. it clears foreign debris from the smell pathways.
   b. it forces air over out-of-the-way olfactory organs.
   c. it stimulates mucous production to dissolve molecules and create "smell".
   d. it narrows the nasal passages to intensify the flow of air.

_____23. The conduction pathway for the sense of smell begins with the
   a. cilia of the olfactory cells in the superior wall of the nasal cavities.
   b. stimulation of the cribriform plate.
   c. olfactory hairs of the receptor cells.
   d. both a and c above are correct.

_____24. The sense of taste is also called
   a. olfaction
   b. gustation
   c. vision
   d. lacrimation

_____25. What is the relationship between taste buds and papillae?
   a. Papillae are located on taste buds.
   b. Taste buds are located within papillae.
   c. Papillae are located on gustatory cells of taste buds.
   d. All of the above are correct.

SPECIAL SENSES-VISION. Much of the material for this section is also covered in the labeling section of this study guide. Read the sections, label the diagrams, then select the best choices for each statement here.

_____26. Which of the following is the conjunctiva most associated with?
   a. photoreceptors
   b. the tear duct
   c. the extrinsic muscles of the eye
   d. the eyelid

_____27. Which of the following is most associated with the lacrimal apparatus?
a. the lacrimal gland
b. the lacrimal sac
c. the nasolacrimal ducts
d. all of the above

_____28. Which muscle of the eye moves the eye laterally?
a. lateral rectus
b. medial rectus
c. superior rectus
d. inferior rectus

_____29. Which muscle of the eye depresses the eye and turns it laterally
a. superior rectus
b. inferior rectus
c. inferior oblique
d. superior oblique

_____30. Which of the following terms is used in conjunction with discussions of the layers of the eye?
a. meninges
b. tunics
c. dermis
d. strata

_____31. The thick, outermost layer of the eyeball containing a sclera and cornea is the
a. fibrous tunic
b. vascular tunic
c. nervous tunic
d. colored tunic

_____32. "Don`t fire until you see the ... of their eyes!"
a. tunic
b. sclera
c. cornea
d. lacrimal

_____33. The transparent "window" of the anterior portion of the eye is the
a. tunic
b. sclera
c. cornea
d. lacrimal

_____34. Which of the following are associated with the vascular tunic of the eye?
   a. sclera and cornea
   b. choroid and ciliary body
   c. iris and lens
   d. both b and c

_____35. "Five foot, two, ... of blue, koochie, koochie, koochie koo." - from a song popular in the "Twenties".
   a. choroid
   b. ciliary body
   c. iris
   d. lens

_____36. The black spot in the center of the eye, which gets larger or smaller depending on the amount of light, is called the
   a. choroid
   b. ciliary body
   c. iris
   d. pupil

_____37. The part of the eye whose loss of transparency is called a cataract is the
   a. choroid
   b. iris
   c. pupil
   d. lens

_____38. Accommodation can be described as
   a. the transparency of the eye
   b. the shape-change of the lens
   c. the pressure in the vitreous humor
   d. a disturbance in the nervous tunic

_____39. Poor drainage of aqueous humor may cause
   a. pain
   b. pressure in the eye
   c. chronic glaucoma
   d. all of the above

_____40. Which type of cell in the retina is most sensitive to low levels of light and would be helpful on a moonlit walk?
   a. bipolar neurons
   b. ganglion cells
   c. rod cells
   d. cone cells

_____41. Which type of cell in the retina is most responsive to well-lit color stimulation?
   a. bipolar neurons
   b. ganglion cells
   c. rod cells
   d. cone cells

_____42. Literally, this word means "rainbow"
   a. choroid
   b. iris
   c. aqueous humor
   d. retina

_____43. Literally, this word means "membrane-like"
   a. choroid
   b. iris
   c. vitreous humor
   d. retina

_____44. Literally, this word means "net"
   a. choroid
   b. iris
   c. aqueous humor
   d. retina

_____45. The "blind spot" is
   a. a place in the retina where there are no photoreceptor cells
   b. the optic disk
   c. the macula lutea
   d. both a and b

_____46. An image focused upon the ... is seen with optimal visual acuity.
   a. optic disk
   b. macula lutea
   c. fovea centralis
   d. photoreceptoralis

_____47. The bending of light rays is called
   a. visual acuity
   b. retraction
   c. refraction
   d. accommodation

_____48. Another term for a normal eye is
   a. emmetropic
   b. myopic
   c. hyperopic
   d. astigmatic

_____49. The condition of being nearsighted is called
   a. emmetropia
   b. myopia
   c. hyperopia
   d. astigmatism

_____50. When an image focuses behind the short eyeball on the retina, and distant objects can still be focused properly, the condition is called
   a. emmetropia
   b. myopia
   c. hyperopia
   d. astigmatism

_____51. What is an astigmatism?
   a. a "lazy" lens
   b. a distorted curvature of the lens
   c. a cloudy lens
   d. a cloudy vitreous humor

_____52. Which of the following is true of the image projected upon the retina?
   a. it is actual sized
   b. it is inverted
   c. it is reversed right to left
   d. both b and c

SPECIAL SENSES - HEARING.

_____ 53. the sense of hearing

_____ 54. the part of your ear that holds earrings

_____ 55. meaning wing-like

_____ 56. drum-like membrane

_____ 57. tube extending into the temporal bone

_____ 58. ear wax

_____ 59. air-filled space of the middle ear

_____ 60. collectively, three small bones of the middle ear

_____ 61. narrow tube between ear and throat

_____ 62. equalizes air pressure on both sides of eardrum

_____ 63. infection in air cells of the mastoid process

_____ 64. ossicle connected to tympanic membrane

_____ 65. hammer

_____ 66. anvil

_____ 67. stirrup

_____ 68. between the stapes and cochlea

_____ 69. inner ear

audition
auditory tube
auditory ossicles
auricle
bony labyrinth
cerumen
cochlea
cristae
dynamic equilibrium
endolymph
Eustachian tube
external auditory canal
incus
labyrinth
malleus
mastoiditis
organ of Corti
otoliths
oval window
perilymph
round window
scala vestibuli
scala tympani
stapes
static equilibrium
tympanic membrane
tympanic cavity
vestibule of the inner ear
vestibule

_____ 70. literally, maze

_____ 71. literally, chamber

_____ 72. literally, snail

_____ 73. "clear fluid around"

_____ 74. "clear fluid within"

_____ 75. series of canals within the temporal bone

_____ 76. fluid between bony labyrinth and membranous labyrinth

_____ 77. fluid inside the membranous labyrinth

_____ 78. upper compartment of cochlear canal

_____ 79. lower compartment of cochlear canal, extends to round window

_____ 80. the organ of hearing

_____ 81. sensation of body position

_____ 82. sensation of rapid movements of the head

_____ 83. calcium carbonate crystals whose movement announces body position

_____ 84. organs of dynamic equilibrium whose movement notifies brain of rapid head movement

audition
auditory tube
auditory ossicles
auricle
bony labyrinth
cerumen
cochlea
cristae
dynamic equilibrium
endolymph
Eustachian tube
external auditory canal
incus
labyrinth
malleus
mastoiditis
organ of Corti
otoliths
oval window
perilymph
round window
scala vestibuli
scala tympani
stapes
static equilibrium
tympanic membrane
tympanic cavity
vestibule of the inner ear
vestibule

INTEGRATIVE FUNCTIONS. To integrate is to process and interpret sensory input before sending out a motor response. Integrative functions include thinking, memory, and emotion. Most of these activities occur in the cerebral cortex. Because the paragraph headings look similar to others in this chapter ("sensory areas, association areas, etc."), read "C. Integrative Functions" in the chapter summary. Also study the illustrations and be able to locate the areas mentioned. Underline the correct phrase or word below.

85. Even with your eyes closed, you can pinpoint skin sensations because of the (general sensory area/somesthetic association area).

86. The lips, tongue, and fingertips contain more receptors than the entire body below the shoulders. (True/False)

87. The (primary visual area/somesthetic association area) interprets impulses from the thalamus and general sensory area.

88. The (primary visual area/primary auditory area) interprets the nature of the sound that is heard.

89. In general, any area of the cerebral cortex not pinpointed as a primary sensory or motor area is considered to have an associative function. (True/False)

90. An example of an association are of the cerebral cortex is the (auditory area/gnostic area).

91. The term (consciousness/coma/stupor) refers to the mental awareness of the brain and ability to respond to stimuli.

92. The cyclic activity of various levels of consciousness in the course of a 24-hour day is called a (comatose consciousness/circadian rhythm).

93. Brain death is determined by a flat baseline on an EEG. (True/False)

94. The motor areas of the cerebral cortex are located primarily in the frontal lobes in front of the central sulcus. (True/False)

95. The functions of thought and memory are performed within the (premotor/association) area of the cerebral cortex.

96. The conscious understanding, or development, of an idea is a (thought/memory)

97. (Short-term/long-term) memory usually lasts from a few second to several hours and deals with small bits of information.

98. There are three types of learning processes. (True/False)

99. Repetition, repetition, repetition stimulates long-term memory. (True/False)

100. The development of short-term memory leads to anatomic changes in the brain. (True/False)

101. The system responsible for our emotions is the (Dhammapada/Limbic).

102. The prefrontal lobotomy is an effective, modern procedure for controlling severely emotionally disturbed patients. (True/False)

MOTOR FUNCTIONS. After reading this entire section, answer the following.

_____103. Motor impulses arise
      a. in the sensory signals
      b. in the integration centers
      c. in the skin or sensory organs
      d. none of the above

_____104. Motor impulses generated in the medulla, pons, and midbrain are regarded as
      a. stimuli
      b. threshold impulses
      c. reflexes
      d. voluntary responses

_____105. The conduction of voluntary motor impulses from motor areas of the brain to skeletal muscles as in the raising of the hand is accomplished by
      a. somatic pathways
      b. reflex arcs
      c. autonomic pathways
      d. voluntary paths

_____106. The upper motor neuron affecting a spinal nerve can be described as
      a. long
      b. short
      c. white
      d. striated

_____107. Which type of neuron carries an impulse to a skeletal muscle effector?
      a. upper motor neuron
      b. lower motor neuron
      c. association neuron
      d. none of the above

CLINICAL TERMS OF THE FUNCTIONAL NERVOUS SYSTEM. After reading this section of text, match the term with its description. Check your work. Correct your spelling. Cover over your answers or erase them, and work through the statements again. Remember, repetition (repetition, repetition) is a key to long term memory!

_____ 108. inability to express thoughts in a connected manner

_____ 109. surgically making small incisions in the cornea to improve myopia

_____ 110. loss of language function caused by blow to the head

_____ 111. progressive loss of brain tissue and memory

_____ 112. total color blindness

_____ 113. dizziness

_____ 114. inflammation of the eustachian tube

_____ 115. general term for mental deterioration

_____ 116. surgical implantation of artificial hearing aid into the cochlea

_____ 117. inability to recognize familiar objects by feeling their shape with the eyes closed

_____ 118. cloudy lens

_____ 119. stroke

acataphasia
achromatopsia
Alzheimer's disease
ametropia
aphasia
astereognosis
cataracts
cerebrovascular accident
cochlear implant
dementia
epilepsy
eustachitis
glaucoma
iridectomy
radial keratotomy
sclerostomy
vertigo

199

_____ 120. surgical procedure to relieve the pressure caused by glaucoma

_____ 121. a defect in the refractive powers of the eye resulting in poor focusing onto the retina

_____ 122. elevated pressure in the eye

_____ 123. episodes of convulsive seizures causes by uncontrolled electrical discharges in the brain

acataphasia
achromatopsia
Alzheimer's disease
ametropia
aphasia
astereognosis
cataracts
cerebrovascular accident
cochlear implant
dementia
epilepsy
eustachitis
glaucoma
iridectomy
radial keratotomy
sclerostomy
vertigo

MASTERY ESSAY

Describe the conversion of sound vibrations into nerve impulses with reference to all of the structures involved.

LABELS AND LISTS

1. Name five types of receptors and the types of stimulus they detect.

   a.

   b.

   c.

   d.

   e.

2. Label the general sensory pathway

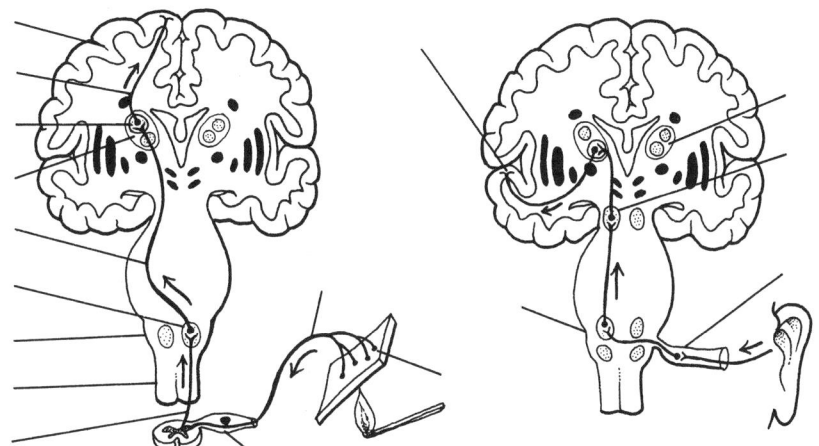

3. Label these structures involved in olfactory sensation.

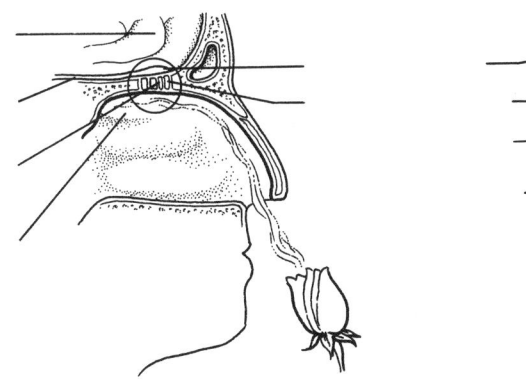
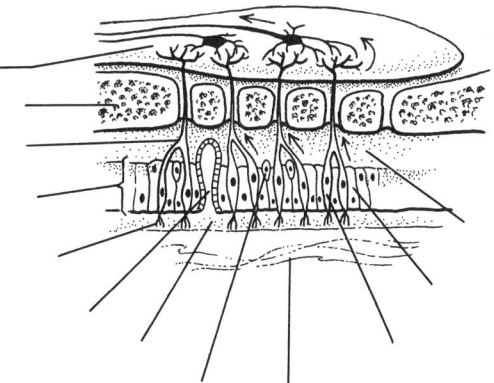

4. Label this drawing of a taste bud.

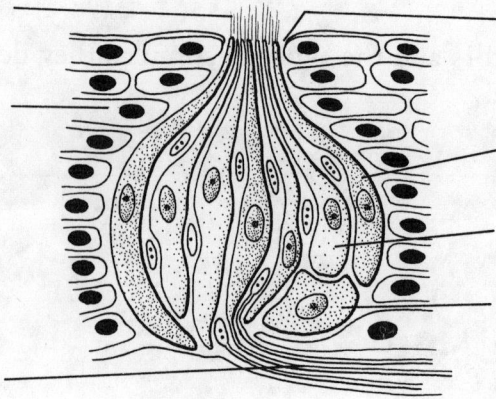

5. Label the regions of the tongue involved in each of the four primary types of taste sensation.

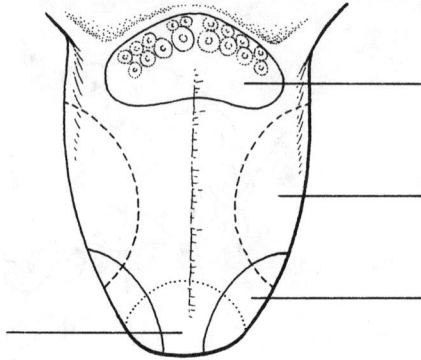

6. Label the components of the lacrimal apparatus.

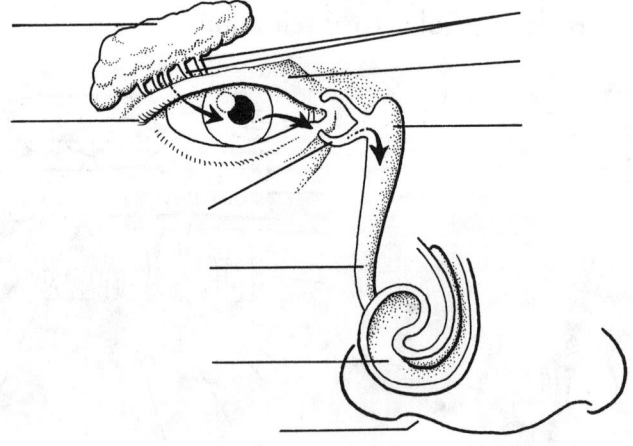

7. Label the eye and its accessory structures.

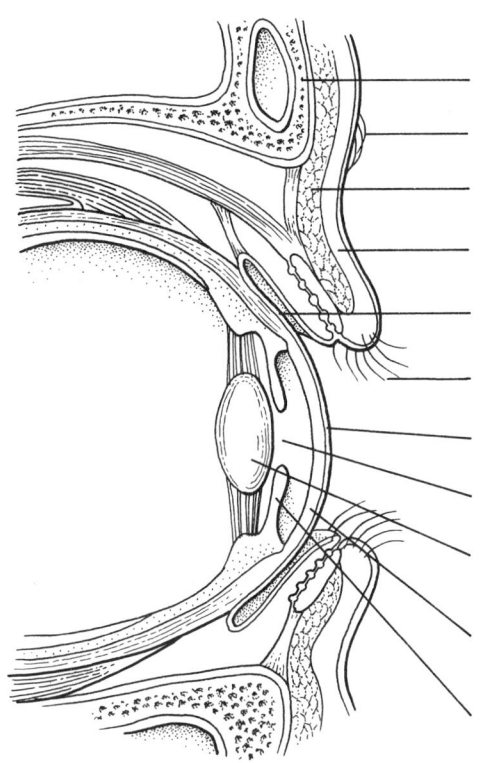

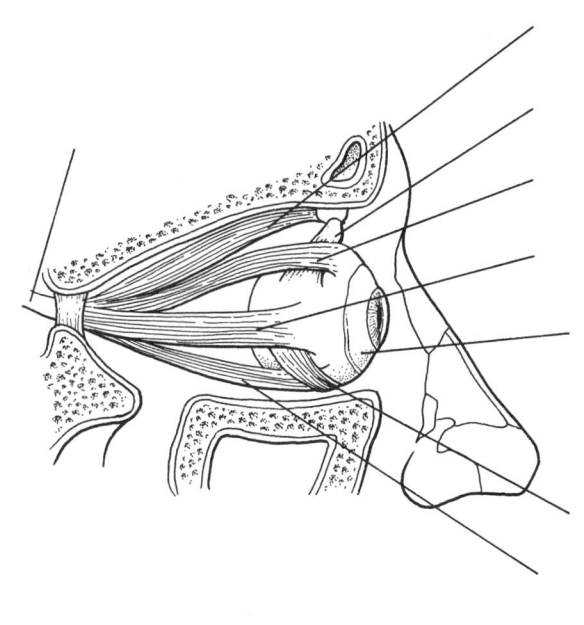
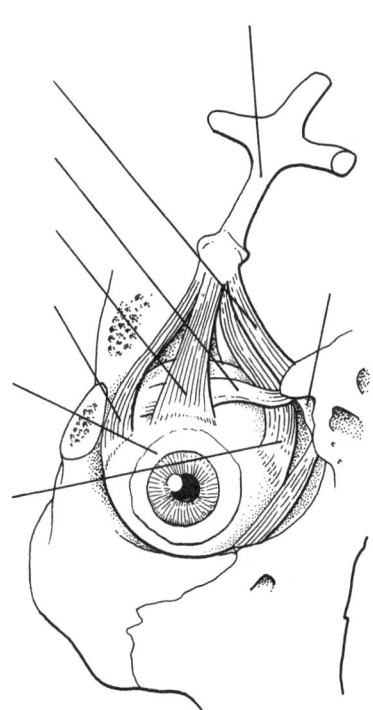

8. Label the eye.

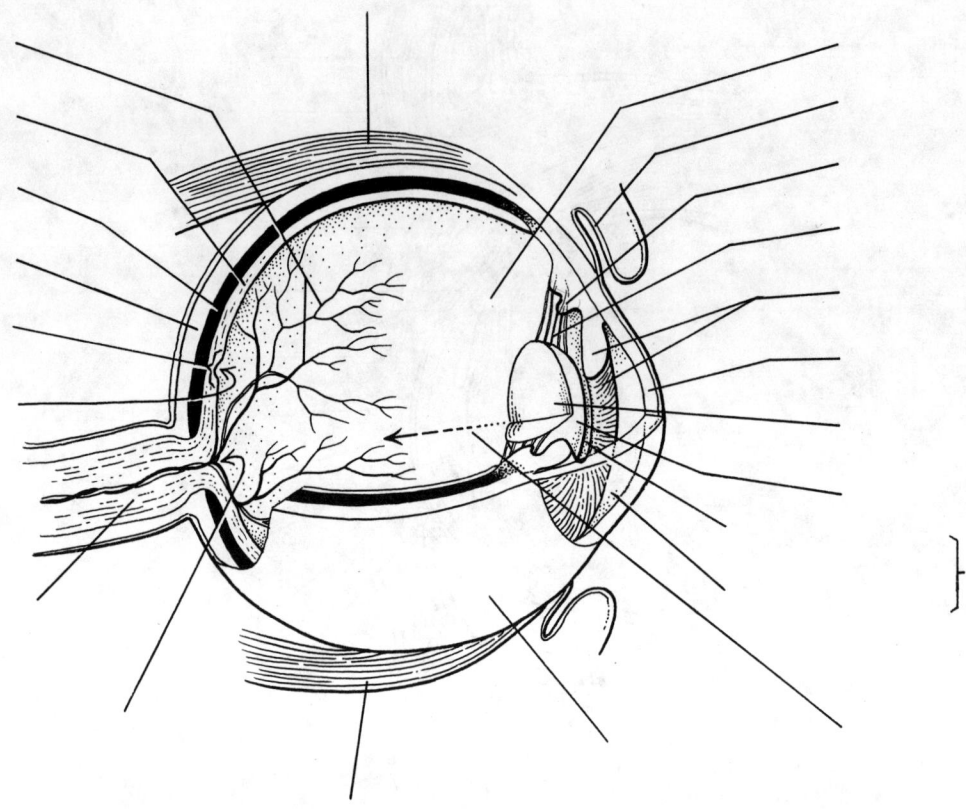

9. Label the ear.

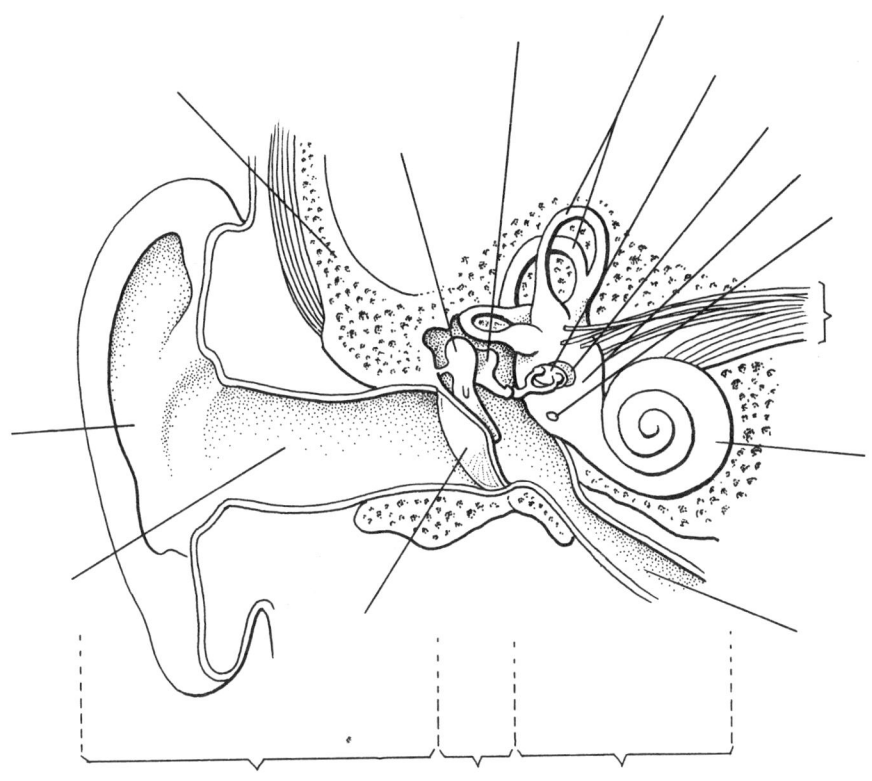

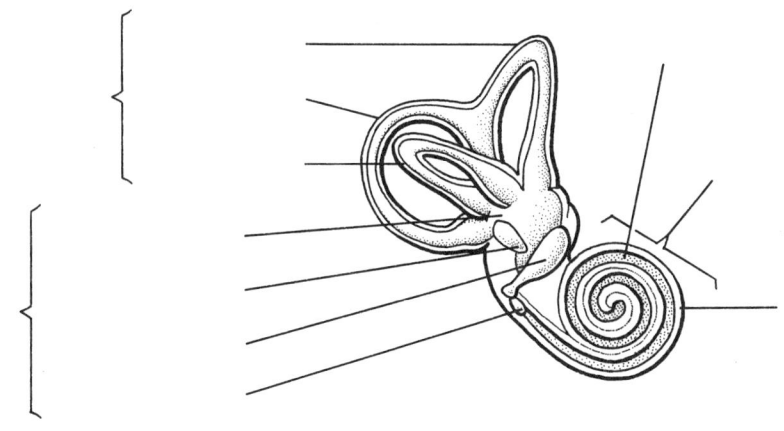

10. Label the structures of the inner ear.

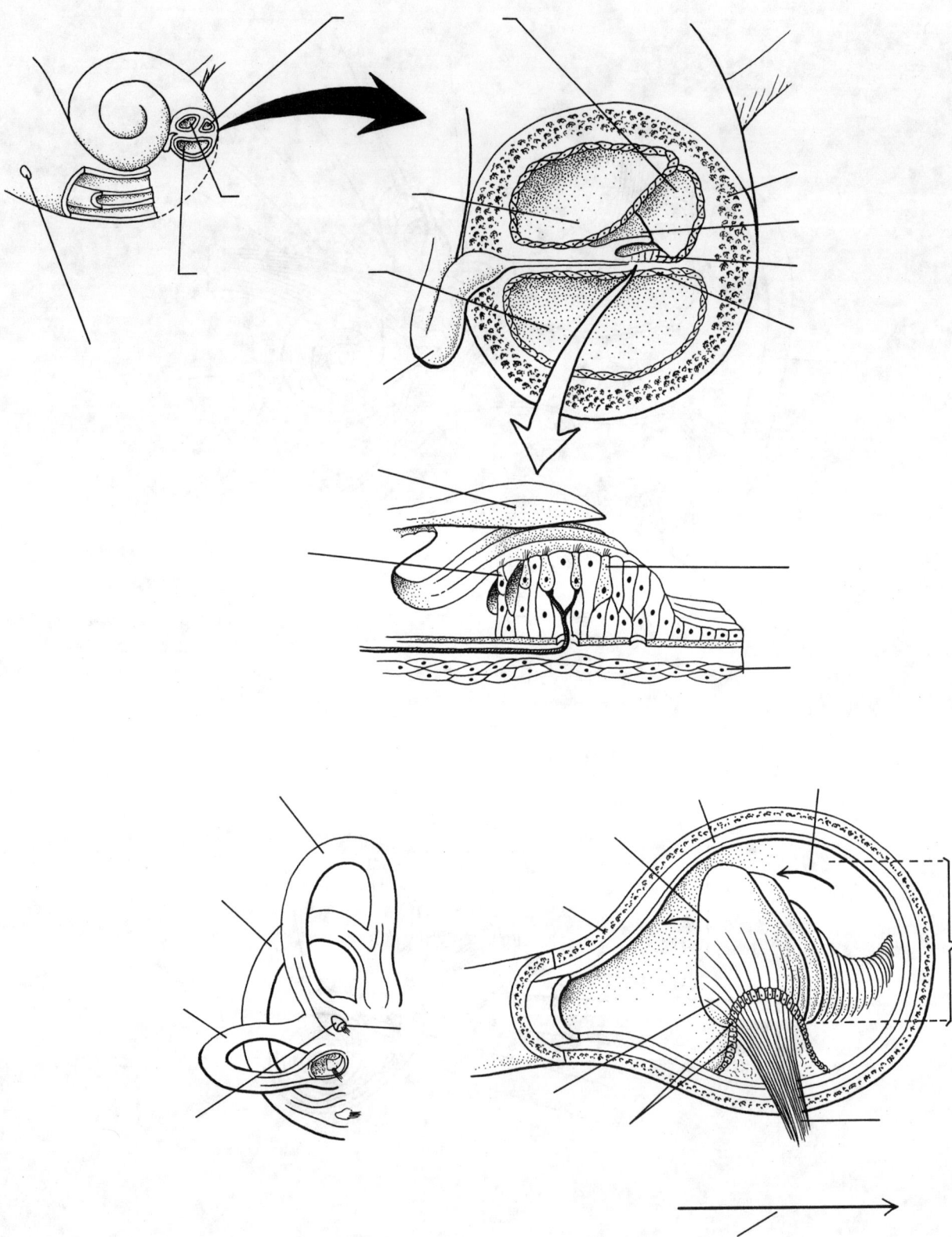

ADDITIONAL STUDY

Review the illustrations in your textbook and understand the answers to the questions associated with each one. The answers are on pages 312-313.

After you have studied the chapter, test yourself by **writing** the answers to the "CONCEPTS CHECKS" and "QUESTIONS FOR REVIEW" in your text. Writing the answers will reinforce your learning, provide confidence for your exams, and help you locate subjects to review.

The day before an exam over this chapter, read the "Learning Objectives", page 280, and review any of the sections which might cause a problem.

CHAPTER TEN REVIEW
Robert W. Bauman, Jr., Ph.D.

- Crosswords Plus

Across

1. interprets nature of sensations
3. involved in static equilibrium
5. first auditory ossicle
7. inner ear
9. proximal taste buds
11. ability to recall the past
13. last auditory ossicle
15. a change which initiates a nerve impulse
18. first cranial nerve
19. lateral taste buds
21. caused by unequal curvature of lens or cornea
24. color sensitive neuron
25. loss of sensation
26. receptor of physical change
27. converts stimulus into impulse
30. organ of vision
32. a function of the occipital lobe
35. transparent fibrous tunic
36. contains the organ of Corti
37. colored portion of eye
40. inability to focau on near objects
41. nerve which carries impulses for hearing
43. distal taste bud
44. state of awareness
45. mental deterioration

Down

1. tatse bud for NaCl
2. eardrum membrane
3. nerve of equilibrium
4. white covering of eye
6. center of emotion
8. nervous tunic
10. processing and interpretation of sensory input
12. cranial nerve of vision
14. receptor of light stimuli
16. inability to focus distant objects
17. episodes of convulsive seizures
20. pigment in rods
22. focusing of lens
23. pain receptors
27. pain felt in a body part removed from its point of origin
28. absorbs light
29. hole
30. fluid inside the membranous labyrinth
31. gland which secretes tears
32. sensation of dizziness
33. conscious understanding
34. sense of taste
38. focuses light
39. middle ear bone
42. contains rhodopsin

208

CHAPTER 11: THE ENDOCRINE SYSTEM

The human body is incredibly designed to function efficiently because of the release of vital chemicals called hormones. The timely production of these chemicals causes growth, mother's milk, many of the differences between men and women, the ability to "leap tall buildings in a single bound", and the life-saving balance of a host of biological molecules. Keep reading for a fascinating and eye-opening study of the endocrine system.

CONTENT MASTERY

COMPOSITION OF THE ENDOCRINE SYSTEM. Read the Introduction and the section dealing with the Composition of the Endocrine System before you answer the following. Fill in the blanks with the most appropriate answer.

_____ 1. The primary role of the endocrine system is in helping to maintain ...

_____ 2. The chemicals released by the endocrine system are known as ...

_____ 3. Compared to the nervous system, the control provided by the endocrine system occurs more (slowly/quickly).

_____ 4. Because the organs of the endocrine system perform mainly a secretory function, they are also called ...

_____ 5. ... glands secrete their hormones into ducts.

_____ 6. ... glands secrete their hormones into the extracellular space around their cells.

_____ 7. Which type of glands makes up the endocrine system?

HORMONES. This section lends itself perfectly to essay questions, so read with a "mental spotlight" looking for good questions (and their answers of course).

_____ 8. chemical units produced by endocrine glands

cAMP
adenylate cyclase
hormones

_____ 9. the type of cell affected by a particular hormone

lipid-soluble hormones
prostaglandin
protein kinases

_____ 10. examples of these types of hormones include NE, ADH, OT, CT, and PTH

target cell
water-soluble hormones

_____ 11. steroid hormones are in this class of hormones

_____ 12. examples of this type of hormones include aldosterone, cortisol, testosterone, and thyroxine

_____ 13. a group of lipids other than hormones that greatly affect various metabolic activities

_____ 14. class of hormones composed of amino acids

_____ 15. class of hormones that are lipid-like

_____ 16. most common second messenger

_____ 17. enzyme which converts ATP into cAMP

_____ 18. enzymes which are activated by a series of reactions triggered by cAMP

_____ 19. enzymes which activate many other proteins within a cell

HORMONAL CONTROL.

20. Describe a negative feedback system, and give an example.

21. Describe a positive feedback system, and give an example.

22. Describe how the nervous system sometimes controls hormone release, and give an example.

THE ENDOCRINE GLANDS - PITUITARY GLAND. As you read about the pituitary gland, fill in this chart describing the hormones secreted and their major effect.

LOBE OF THE PITUITARY GLAND	HORMONE SECRETED	PRIMARY EFFECT
anterior	growth hormone	23.
anterior	24.	increases skin pigmentation
anterior	25.	stimulates milk secretion
anterior	26.	stimulates adrenal cortex
anterior	Thyroid-stimulating hormone	27.
anterior	Follicle-stimulating hormone	28.
anterior	29.	stimulates ovulation in females and production of testosterone in males
posterior	30.	uterine contractions and milk release
posterior	Antidiuretic hormone	31.

32. How does the hypothalamus affect the hormone secretions of the pituitary gland?

33. Differentiate between tropic and gonadotropic hormones.

Match the following pituitary hormones with the appropriate statements. Terms may be used more than once or not at all. Writing your answers in the blanks should help you learn to spell the hormones and become familiar with them. Once you know something's "name", you feel better about studying about it.

_____ 34. hormone from the anterior lobe of pituitary that helps maintain a relatively constant blood sugar level

_____ 35. condition of low blood sugar levels

_____ 36. condition of high blood sugar levels

_____ 37. caused by excessive production of GH in a youngster

_____ 38. caused by excessive production of GH in an adult

_____ 39. caused by insufficient production of GH during childhood

_____ 40. actually produces the hormones stored in the posterior lobe of the pituitary

acromegaly
adrenocorticotropic hormone
antidiuretic hormone
follicle-stimulating hormone
gigantism
growth hormone
hypoglycemia
hyperglycemia
hypothalamus
luteinizing hormone
osmoreceptors
oxytocin
pituitary dwarfism
prolactin
thyroid-stimulating hormone

_____ 41. literally, a neutral substance, swift birth

_____ 42. helps the body conserve water

_____ 43. Increase in osmotic pressure causes these to trigger ADH production.

_____ 44. hormone sometimes referred to as a vasopressin

_____ 45. hormone which helps regulate blood pressure

**acromegaly
adrenocorticotropic hormone
antidiuretic hormone
follicle-stimulating hormone
gigantism
growth hormone
hypoglycemia
hyperglycemia
hypothalamus
luteinizing hormone
osmoreceptors
oxytocin
pituitary dwarfism
prolactin
thyroid-stimulating hormone**

Write the correctly spelled pituitary hormone that corresponds to each of the abbreviations given.

46. GH _____

47. MSH _____

48. PRL _____

49. ACTH _____

50. TSH _____

51. FSH _____

52. LH _____

53. OT _____

54. ADH _____

THYROID AND PARATHYROID GLANDS. Locate these glands on a diagram in your text. As you read about these glands and their hormones, carefully learn to pronounce the hormones. Be sure you pronounce them aloud each time you read their names. Do not simply "read over" their names or their abbreviations. Get "fluent" in the language of the endocrine system!

Fill in the following chart describing the glands, hormones, and their primary effects.

GLAND	HORMONE SECRETED	PRIMARY EFFECT
Thyroid	Thyroxin, Triiodothyronine	55.
Thyroid	56.	reduces Ca and P levels in the blood
57.	Parathyroid hormone	increases Ca levels in the blood and decreases phosphate level in the blood

Fill in the blanks concerning the thyroid and parathyroid glands.

_____ 58. Thyroxine and Triiodothyronine are given the abbreviations T4 and T3, respectively, because of the number of atoms of ... in each of their molecules.

_____ 59. Both T4 and T3 stimulate the rate of (59), promote (60) synthesis, increase the rate of (61) uptake,

_____ 60. promote (62) metabolism, and accelerate actions of the (63) system!

_____ 61.

_____ 62.

_____ 63.

_____ 64. ... reduces the calcium and phosphate levels in the blood.

_____ 65. As calcium levels rise, calcitonin secretion (increases/decreases).

_____ 66. The parathyroid hormone increases blood calcium levels in opposition to the effects of ... released by the thyroid gland.

_____ 67. When calcium levels in the blood drop 40% below normal, nerve fibers become ... (which may result in convulsions, tetany, and possibly death).

ADRENAL GLANDS. Locate and label the adrenal glands on the figure. Then, as you read through this section, fill in the following chart.

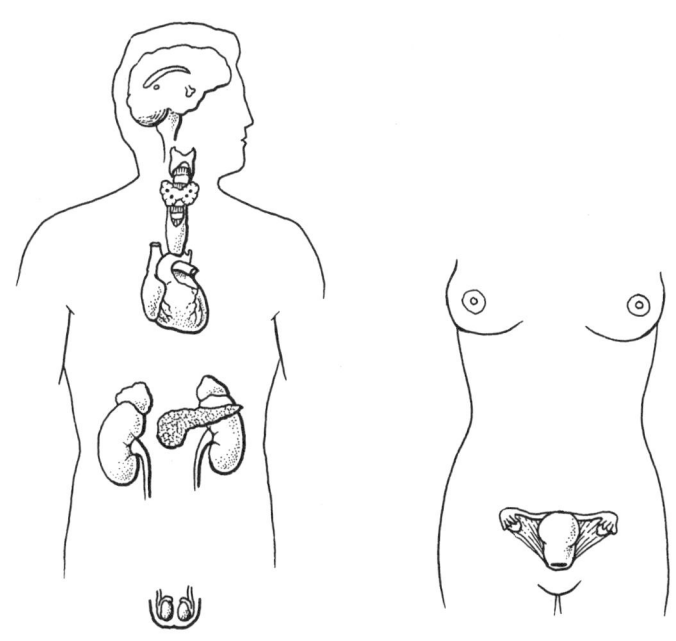

GLAND	HORMONE SECRETED	PRIMARY EFFECT
Adrenal medulla	68.	prolongs "fight or flight", speeds heart metabolism, blood pressure, etc.
69.	Mineralcorticoids	70.
Adrenal cortex	71.	stimulates glycogen formation and storage, reduces inflammation
Adrenal cortex	androgens and estrogens	72.

Select the answer which best completes each of the following concerning the adrenal glands.

_____73. The hormones secreted by the adrenal medulla include
   a. epinephrine
   b. norepinephrine
   c. aldosterone
   d. both a and b

_____74. Which of the following is a true statement?
   a. Norepinephrine is the same substance released by parasympathetic neurons.
   b. Epinephrine has a "fight" effect while norepinephrine has a "flight" effect on the body.
   c. Epinephrine is produced from norepinephrine by enzymes within secretory cells.
   d. Epinephrine and norepinephrine have the same effect as a neurotransmitter though their effects do not last as long as those of the neurotransmitter.

_____75. The emergency gland of the body
   a. is the adrenal medulla
   b. is the adrenal cortex
   c. allows quick thinking and fast running
   d. both a and c

_____76. The secretion of epinephrine and norepinephrine is increased by
   a. the hypothalamus
   b. anxiety
   c. the anterior lobe of the pituitary
   d. anxiety stimulates the hypothalamus which signals the production of epinephrine and norepinephrine

_____77. The mineralcorticoids, glucocorticoids, and sex hormones produced by the adrenal cortex are synthesized from what "forbidden" dietary molecule?
   a. sugar
   b. fat
   c. cholesterol
   d. caffeine

_____78. The primary mineralocorticoid is
   a. glucagon
   b. aldosterone
   c. cortisol
   d. corticosterone

_____79. Aldosterone causes cells in various areas to retain... and remove...
   a. cholesterol, fat
   b. potassium, sodium
   c. calcium, sodium
   d. sodium, potassium

_____80. An indirect effect of aldosterone secretion is
   a. an increase in blood sugar
   b. a decrease in blood sugar
   c. an increase in blood pressure
   d. a decrease in blood pressure

_____81. Which of the following is NOT a glucocorticoid?
   a. cortisol
   b. corticosterone
   c. cortisone
   d. gluconeogenesis

_____82. The primary role of glucocorticoids is
   a. to promote glucose production
   b. to promote gluconeogenesis
   c. to promote glycogen synthesis
   d. all of the above

_____83. Which hormone group inhibits allergic responses?
   a. mineralcorticoids
   b. renin-angiotensin
   c. androgens
   d. glucocorticoids

_____84. The hormones which have a masculinizing effect on both men and women are
   a. androgens
   b. estrogens

_____85. The hormones which have a feminizing effect on both men and women are
   a. androgens
   b. estrogens

_____86. Both androgens and estrogens are produced in the male.
   a. true, but more androgens than estrogens
   b. true, but more estrogens than androgens
   c. false

THE PANCREAS. Match these terms with the correct statement or definition. Terms may be used more than once or not at all.

_____ 87. located in the abdominal cavity behind the stomach

_____ 88. two hormones secreted by the pancreas

_____ 89. endocrine cells of the pancreas

_____ 90. type of cell which secretes glucagon

_____ 91. type of cell which secretes insulin

_____ 92. pancreatic hormone stimulating the conversion of glycogen into glucose

_____ 93. pancreatic hormone stimulating the formation of glycogen from glucose

_____ 94. literally, presence of glucose

_____ 95. literally, a neutral substance in little islands

_____ 96. disease of the islets of Langerhans during adolescence

_____ 97. disease of the islets of Langerhans which develops in the "fat and forty" group

alpha cells
beta cells
glucagon
insulin
islets of Langerhans
pancreas
Type I diabetes mellitus
Type II diabetes mellitus

THE GONADS, PINEAL GLAND, AND THYMUS GLAND. Match these terms with the correct statement or definition. Terms may be used more than once or not at all.

_____ 98. organs that produce sex cells and secrete primary sex hormones

_____ 99. female gonads

_____ 100. male gonads

_____ 101. female hormones produced by female gonads (two answers)

_____ 102. male hormone

_____ 103. sometimes called the epithalamus

_____ 104. probably secretes melatonin

_____ 105. a gland which diminishes with age

_____ 106. hormone secreted by thymus gland

_____ 107. gland which plays a role in the immune system

_____ 108. hormone which stimulates the production of T-lymphocytes

_____ 109. endocrine gland located nearest the heart

estrogens
gonads
melatonin
ovaries
pineal gland
progesterone
testes
testosterone
thymus gland
thymosin

HOMEOSTASIS. As is typical of a discussion of homeostasis, this section lends itself wonderfully to essay questions!

110. Contrast the nervous system and endocrine system in terms of their maintenance of homeostasis.

111. Discuss the importance of a balanced output of hormones by the endocrine system mentioning the consequence of hypersecretion and hyposecretion. Choose one example to illustrate your answer.

112. Study Table 11-2 in your text. Cover various columns to test your memory. Create a blank table of your own to test your memory. Have a parent, study partner, spouse, or other friend quiz your memory.

CLINICAL TERMS OF THE ENDOCRINE SYSTEM.

_____ 113. caused by a deficiency in ADH

_____ 114. too much cortisol from the adrenal cortex resulting in fat accumulation, weak muscles, easy bruising

_____ 115. caused by a decreased production of aldosterone and cortisol

_____ 116. too much prolactin, no more menstrual periods

_____ 117. hypothyroidism in children

_____ 118. hypothyroidism in adults

_____ 119. hyperthyroidism, weight loss, nervousness, hypertension

Addison's disease
aldosteronism
amenorrhea
cretinism
Cushing's syndrome
diabetes insipidus
diabetes mellitus
exophthalmos
goiter
Graves' disease
myxedema

_____ 120. protrusion of the eyes due to hyperthyroidism

_____ 121. caused by insufficient insulin

_____ 122. enlargement of thyroid gland, can be caused by lack of iodine in the diet

_____ 123. can be caused by prolonged stress or obesity

_____ 124. can cause the neck to swell

_____ 125. associated with too much sodium in the body resulting in high blood pressure

_____ 126. associated with too little sodium in the body resulting in dehydration

**Addison's disease
aldosteronism
amenorrhea
cretinism
Cushing's syndrome
diabetes insipidus
diabetes mellitus
exophthalmos
goiter
Graves' disease
myxedema**

LABELS AND LISTS

1. List the six primary endocrine glands of the body.

   a.   d.

   b.   e.

   c.   f.

2. What two endocrine glands provide a minor role in homeostasis?

   a.   b.

3. What organs provide hormones in addition to their main functions in the body?

   a.   b.

   c.   d.

221

4. Mention four ways that a hormone can alter a cell's metabolic processes.

   a.   b.

   c.   d.

5. List 8 endocrine glands.

   a.   e.

   b.   f.

   c.   g.

   d.   h.

6. List 9 hormones released by the pituitary.

   a.   f.

   b.   g.

   c.   h.

   d.   i.

   e.

7. List the three primary hormones produced by the thyroid gland.

   a.   b.

   c.

8. List three classes of compounds secreted by the adrenal cortex.

   a.   b.

   c.

ADDITIONAL STUDY

The figures in your textbook illustrate many features of endocrine system. Review each figure and answer the questions associated with each one. Also answer the questions given throughout the chapter as "Concept Checks" (pages 316, 320, 321, 326, 328, 332, and 334). The correct answers to the art legend questions are given on pages 341-342.

After you have studied the chapter, test yourself by **writing** the answers to the "QUESTIONS FOR REVIEW" on pages 340 and 341 in your textbook. Writing the answers will reinforce your learning.

The day before an exam over this chapter, read the "Learning Objectives", page 314, and review any of the sections which might cause a problem for you.

ENDOCRINE SYSTEM
Robert W. Bauman, Jr., Ph.D.

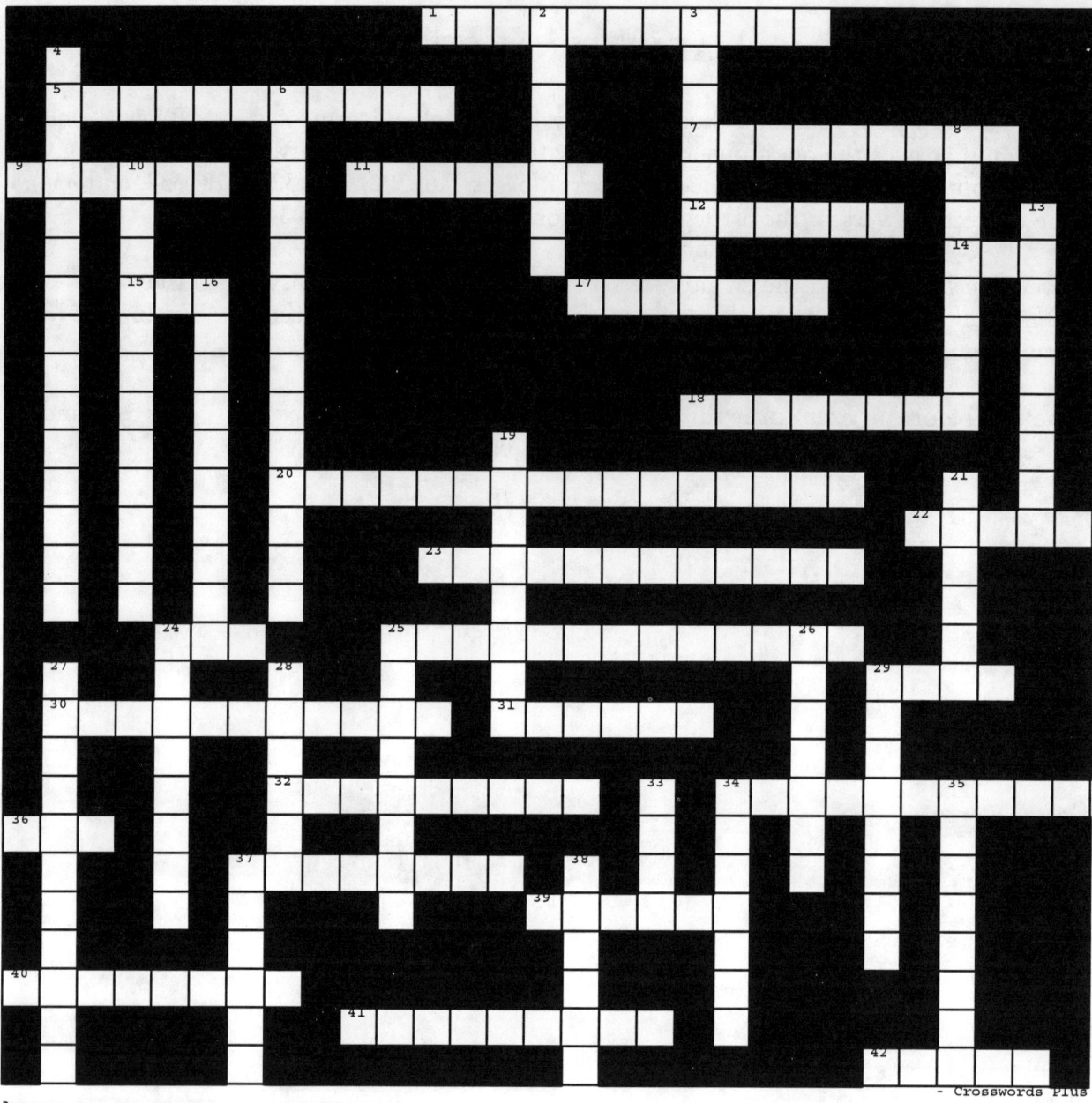

Across

1. four glands
5. sympathomimetic hormone
7. ductless glands
9. enlarged thyroid
11. chemical secreted by one body part which affects another body part
12. gland imprtant in development of T lymphocytes
14. chemical which causes conservation of water
15. stimulates follicular cells of the thyroid
17. two-lobed gland in the neck
18. secretes into ducts
20. contains three atoms of iodine
22. hormones composed of amino acids are soluble in ...
23. reducing urine output
24. stimulates the development of ova
25. 'local hormone'
29. controls secretion of hormones from the cortex of the adrenal gland
30. stimulating the release of an ovum
31. steroid hormones are solble in ...
32. female hormones
34. reduces calcium and phosphate levels in the blood
36. causes darkening of the skin
37. causes catabolism of glycogen
39. hormone which stimulates an endocrine gland
40. type of feedback seen with uterine contractions
41. stimulates milk secretion
42. sex organ

Down

2. triangular gland superior to kidney
3. protein molecule which recognizes hormone
4. type of system using cAMP
6. over secretion
8. typical feedback
10. male hormone
13. thyroid hormone
16. pituitary
19. glucocorticoid
21. cell reacting to a hormone
24. process whereby secretions are controlled
25. endocrine and exocrine gland
26. reduces glucose level
27. mineralcorticoid
28. secretes melatonin
29. homeostatic imbalance due to hyposecretion of aldosterone
33. second messenger
34. controlled by calcitonin and PTH
35. stimulates labor
37. caused by hypersecretion of thyroid
38. hypersecretion of this hormone in adults results in acromegaly

- Crosswords Plus

CHAPTER 12: THE BLOOD

Was "blood" one of the first things you wanted to view with a microscope? For many, there is a fascination with this life-sustaining tissue -- the only liquid tissue in the body.

CONTENT MASTERY

FUNCTIONS AND PROPERTIES OF BLOOD. As you read the Introduction and these two sections, fill in the following. This will help pace your reading.

_____ 1.
_____ 2.
_____ 3.
_____ 4.
_____ 5.
_____ 6.
_____ 7.
_____ 8.
_____ 9.
_____ 10.
_____ 11.
_____ 12.
_____ 13.
_____ 14.
_____ 15.
_____ 16.
_____ 17.

Blood can be studied as part of the (1), which includes the (2) system and the (3) system. Blood may also be studies as part of the (4) system which includes red bone marrow, the lymph nodes, and the spleen. In blood, the cellular elements are known as (5) while the watery fluid is called (6).

Three homeostatic functions of blood are (7), (8), and (9). Blood protects the body from bacteria by white blood cells and specialized proteins called (10).

Human blood has certain characteristics which can indicate health. These include color, volume, (11) and (12). The red color of blood is due to a protein called (13). Venous blood is blue (14 true/false). About (15) percent of total body weight is blood. The condition of excessively acidic blood is called (16). Too little acid in the blood is called (17).

PLASMA. Read (with eyes and brain) about plasma, and the substances in this liquid portion of blood. Then answer the following true/false questions. Correct the false statements by writing a word in the blank which will correctly replace the italicized word.

_____18. Plasma is the *liquid* portion of blood.

_____19. Plasma is a transport medium for *hormones*.

_____20. Plasma is mainly composed of *solutes*.

_____21. *Albumins* constitute about 55% of the plasma proteins.

_____22. Albumins *decrease* blood's viscosity.

_____23. Gamma globulins serve as *antigens* for the body's defense.

_____24. Fibrinogen and fibrin are important in the *immune* response.

_____25. Plasma transports nonprotein *nitrogenous* substances such as amino acids and uric acid.

_____26. The gases that are dissolved in plasma include *oxygen* and carbon dioxide.

_____27. Nutrients are carried in *plasma*.

_____28. The most abundant electrolyte dissolved in plasma is *chloride*.

FORMED ELEMENTS. Match these terms with their descriptions.

_____ 29. Besides cells, these formed elements are found in the blood.	**blood smear**
_____ 30. another term for red blood cells	**erythrocytes**
_____ 31. another term for white blood cells	**hematopoiesis**
_____ 32. another term for platelets	**hemocytoblasts**
_____ 33. a procedure using a centrifuge to separate blood cells in a sample of blood	**hematocrit**
_____ 34. a procedure involving a stain for viewing blood under the microscope	**leukocytes**
_____ 35. process of making the formed elements of blood	**red bone marrow**
_____ 36. primary site of hematopoiesis	**stem cells**
_____ 37. original cells from which the formed elements of blood originate (two terms)	**thrombocytes**

ERYTHROCYTES. Try reading these few pages using a very successful study technique: Crease a sheet of notebook paper in half vertically. As you read, write terms or fragments of questions on the left. On the right, define the term briefly or answer the question. In this manner, you will transfer all pertinent information to a form that can be used to quiz yourself. When you complete this exercise, answer the following multiple choice questions. Then, go back over your study sheet. Without looking at the definitions, write a definition for each term. Finally, write the terms while looking only at the definitions.

_____38. The most abundant type of formed element in the blood are
   a. erythrocytes
   b. leukocytes
   c. platelets
   d. stem cells

_____39. Over what percentage of the total blood volume is attributed to red blood cells?
   a. 95
   b. 10
   c. 40
   d. 50

_____40. Which of the following statements is NOT TRUE concerning anemia?
   a. Pernicious anemia can be caused by shortage of folic acid and B_{12}.
   b. Pernicious anemia causes a sleepy, droopy feeling.
   c. Hemorrhagic anemia is due to excessive bleeding and is always fatal.
   d. In hemolytic anemia, the red blood cells are destroyed by phagocytes.

_____41. Sickle cell anemia
   a. is a disease common in the former Soviet Union.
   b. is caused by an inherited inability to make normal hemoglobin molecules.
   c. is effectively treated with blood transfusions or oxygen supplements.
   d. is caused by a lack of intrinsic factor noticed in the synthesis of red blood cells.

_____42. The structure of an erythrocyte can be described as
   a. small
   b. flexible
   c. biconcave
   d. all of the above

_____43. Mature erythrocytes do not have nuclei. What advantages have been suggested?
   a. They do not need nuclei.
   b. The space is better used for hemoglobin.
   c. Loss of nuclei leaves the cell biconcave which enlarges the surface area for the diffusion of gases.
   d. All of the above are suggested.

_____44. The name "erythrocyte" means
   a. oxygen cell
   b. red cell
   c. small cell
   d. disk cell

_____45. Hemoglobin
   a. is a pigment
   b. is a protein
   c. carries a pigment
   d. both a and b

_____46. The chemical binding of iron in hemoglobin with oxygen
   a. causes the erythrocyte to become bright red
   b. causes the erythrocyte to become deep, bluish red
   c. does not cause a change in the appearance of the erythrocyte
   d. none of the above

_____47. In the adult, erythrocytes are produced
   a. in the liver
   b. in the spleen
   c. at three distinct locations
   d. exclusively within red marrow in bones

_____48. The erythrocyte lives
   a. for years
   b. the entire lifetime of the individual
   c. about 4 months
   d. about 1 year

_____49. The hormone which stimulates the production of red blood cells in the marrow is
   a. thyroxin
   b. erythrocytonin
   c. erythropoietin
   d. biliverdin

_____50. Which of the following is an effect of macrophages?
   a. They "eat" damaged erythrocytes.
   b. They wander throughout the body to find old erythrocytes.
   c. They help recycle erythrocytes.
   d. All of these are correct.

_____51. Which is NOT a result of erythrocyte recycling?
   a. macrophages are phagocytized at the rate of 10 billion per hour.
   b. Biliverdin is released to the liver.
   c. Hemoglobin is broken down to a greenish then orange pigment.
   d. The iron is returned to the red marrow or stored in the spleen.

_____52. Which of these foods contain B-complex vitamins and are thus necessary in the diet for the production of erythrocytes? (You may have more than one answer.)
   a. green peas
   b. granola
   c. squash
   d. tomatoes

_____53. Deficiency of folic acid in the diet during the first two weeks of pregnancy causes
   a. tiredness
   b. spina bifida in the infant
   c. a form of diabetes called infantile diabetes
   d. folinemia

LEUKOCYTES. As you read about the white blood cells, answer the following. In most cases, your answers will be one or two words.

_____ 54. What does "leukocyte" literally mean?

_____ 55. What is the general function of a leukocyte?

_____ 56. Do mature leukocytes contain a nucleus?

_____ 57. Into what two groups can leukocytes be put?

_____ 58. What is the most abundant granulocyte in the blood?

_____ 59. Which granulocyte stains red in eosin stain?

_____ 60. Which granulocyte type is rarest and has an s-shaped nucleus?

_____ 61. Name two types of agranulocytes.

_____ 62. Which leukocyte's name literally means neutral lover?

_____ 63. Which type of leukocyte's name literally means rose-colored lover?

_____ 64. Which type of leukocyte's name literally means basic lover?

_____ 65. Which is the smallest type of leukocyte?

_____ 66. Which type of agranulocyte is the largest?

_____ 67. When leukocytes move out of a blood vessel, do they move through the cells of the vessel, through pores in the vessel wall, or between the cells of the vessel?

_____ 68. What word describes the process of leukocytes leaving blood vessels to engulf problem cells?

_____ 69. How are leukocytes attracted to problem areas?

_____ 70. Describe the composition of pus.

_____ 71. Some leukocytes fight germs by phagocytosis. What kind of proteins are produced by lymphocytes to fight disease?

_____ 72. What substance produced by basophils causes inflammation?

_____ 73. Which two types of leukocytes are most specialized for phagocytosis?

_____ 74. What is a good purpose for edema?

_____ 75. What drugs can be taken to reduce inflammation in cases when edema is harmful to healing?

PLATELETS. Indicate which of the following statements are true. If the statement is false, correct the italicized word to make the sentence true.

_____ 76. Another terms for platelets is *thrombocytosis*.

_____ 77. Platelets *have* a nucleus.

_____ 78. Platelets are *larger* than red blood cells.

_____ 79. Platelet literally means *clot* cell.

_____ 80. Platelets are important in the formation of *blood* clots.

HEMOSTASIS. Did you read "homeostasis" instead of "hemostasis"? They are different. **Hemostasis** is a three-step process that stops bleeding so that **homeostasis** is maintained!

_____ 81.	body's mechanism for the stoppage of bleeding	atherosclerosis
		blood vessel spasm
		blood clot
_____ 82.	smooth muscle response to a cut in blood vessel wall	coagulation
		collagen
		embolus
_____ 83.	substance released by platelets which prolongs blood vessel muscle contraction	fibrin
		fibrinolysis
		hemostasis
		homeostasis
_____ 84.	literally, blood stopping	platelet plug
		thrombus
_____ 85.	sticky platelets at broken vessels adhere to these fibers	
_____ 86.	clump of platelets and collagen	
_____ 87.	result of coagulation	
_____ 88.	long threads of protein formed from fibrinogen	
_____ 89.	process in which enzymes are activated to digest fibrin of a blood clot	
_____ 90.	disease in which unwanted blood clot may form along fatty deposits in vessel wall	
_____ 91.	undesirable blood clot	
_____ 92.	free-floating thrombus	

93. Sequence the following events in the process of coagulation:
A. fibrinogen converted to fibrin by enzyme action of thrombin
B. formed elements trapped in net of fibrin
C. fibrin sticks to blood vessels
D. thromboplastin converts prothrombin into thrombin
E. thromboplastin released by platelets and blood vessel walls

_____ _____ _____ _____ _____

BLOOD GROUPS. Read this section. When you recognize a testable statement, reread that statement aloud. Then, answer the multiple choice questions that follow.

_____ 94. Agglutination is
   a. literally, round hole process
   b. the clotting of blood cells
   c. the clumping of blood cells
   d. the plugging of blood vessels

_____ 95. What is the basis for blood grouping?
   a. ethnic origin
   b. surface proteins on RBCs
   c. antigens and antibodies
   d. both b and c

_____ 96. Particular proteins located on the surface of erythrocytes used in blood grouping
   a. antigens
   b. antibodies
   c. agglutinogens
   d. agglutinins
   e. a and c

_____ 97. Proteins within the plasma which are used in blood grouping are
   a. antigens
   b. antibodies
   c. agglutinogens
   d. agglutinins
   e. b and d

_____ 98. The process of identifying the antigens present on the RBC plasma membrane
   a. agglutination
   b. blood grouping
   c. blood typing
   d. blood clotting

_____ 99. A person with blood type AB has
   a. antigen A
   b. antigen B
   c. antigens A and B
   d. antigen O

_____100. A person with blood type O has
   a. antigen A
   b. antigen B
   c. antigens A and B
   d. neither antigen A nor antigen B

_____101. A person with type B blood has
   a. anti-A antibodies
   b. anti-B antibodies
   c. no antibodies
   d. both anti-A and anti-B antibodies

_____102. A person with type AB blood
   a. can donate blood to a person with type O
   b. can receive blood from types A, B, AB, and O
   c. can donate blood to a person with type A
   d. can donate blood to a person with type B

_____103. A person with Rh antigens would have
   a. Rh - blood
   b. both anti-A and anti-B antibodies
   c. Rh+ blood

Fill in the following chart. You should understand this material so well that you could start with a blank sheet of paper and write your own chart identical to this one!

BLOOD TYPE	ANTIGENS ON RBC	ANTIBODIES IN PLASMA	CAN DONATE TO	CAN RECEIVE FROM
A	104.	anti-B	105.	A, O
B	106.	107.	108.	B, O
AB	A & B	109.	110.	111.
O	112.	anti-A & anti-B	A, B, AB, O	113.

Note 1: What is the relationship between agglutination, clumping, coagulation, and clotting? Agglutination is the same thing as clumping. This is an artificial situation resulting from mixing incompatible blood types, or a blood type with its opposite antibody. Coagulation, on the other hand, is the naturally occurring formation of a blood clot.

Note 2: Of prime importance in blood donation are the antibodies present in the **recipient's** blood. Therefore, the small amount of anti-A antibody in type O blood does not prevent donation from type O to type A, because it is diffused by the plasma.

HOMEOSTASIS AND THE BLOOD. Read this section out loud, as if you were giving a guest presentation to a group of pre-medical students. Be convincing. Then answer the following questions.

114. What are the three major functions of the blood?

115. How are the blood's transportation abilities important to homeostasis?

116. How does the blood's protective function maintain homeostasis?

117. How are the blood's regulatory mechanisms important to homeostasis?

CLINICAL TERMS OF THE BLOOD. Match these terms with their descriptions. Terms may be used more than once or not at all. Pay attention to spelling as you write each term. When you correct your answers, go back through the questions and cover the answers, as if they were fill-in-the blank questions with no words from which to choose.

_____ 118. loss of blood due to profuse bleeding

_____ 119. low number of erythrocytes due to shortage of B vitamins

_____ 120. low number of erythrocytes due to macrophage action

_____ 121. decreased blood volume due to blood loss

_____ 122. critical reduction of blood to vital tissues because of heart failure

_____ 123. chronic type anemia with sickle-shaped RBCs

_____ 124. cancer of the blood

_____ 125. hereditary disorder of coagulation process

_____ 126. acute infection of bloodstream by bacteria

_____ 127. oxygen deficiency in the blood

_____ 128. bloodstream infection by *Plasmodium*

_____ 129. inherited, chronic type of hemolytic anemia with dysfunctional hemoglobin

anoxia
apheresis
bacteremia
cyanosis
direct transfusion
exchange transfusion
hemolytic anemia
hemophilia
hemorrhage
hemorrhagic anemia
heparinized whole blood
indirect transfusion
leukemia
leukocytosis
leukopenia
malaria
pernicious anemia
polycythemia
septicemia
shock
sickle cell anemia
thalassemia

_____ 130. persistent "blood-poisoning"

_____ 131. elevated leukocyte count

_____ 132. lowered leukocyte count

_____ 133. bluish skin coloration due to oxygen deficiency in blood

_____ 134. transfer of blood directly from one person to another without exposure of the blood to air

_____ 135. medical technique for cleansing the blood

_____ 136. an abnormal increase in the number of erythrocytes accompanied by thicker blood and higher blood pressure

_____ 137. can be caused by some drugs which destroy white blood cells

anoxia
apheresis
bacteremia
cyanosis
direct transfusion
exchange transfusion
hemolytic anemia
hemophilia
hemorrhage
hemorrhagic anemia
heparinized whole blood
indirect transfusion
leukemia
leukocytosis
leukopenia
malaria
pernicious anemia
polycythemia
septicemia
shock
sickle cell anemia
thalassemia

LABELS AND LISTS

1. What are the three categories of formed elements in the blood? Give the amount of each found in the blood.

    a.

    b.

    c.

2. Name three types of anemia (disease conditions that involve a reduction in the number or efficiency of erythrocytes).

    a.

    b.

    c.

3. Name the two main groups of leukocytes.

    a.

    b.

4. List three types of granulocytes.

    a.

    b.

    c.

5. List the types of leukocytes and the percentage each is of the total leukocyte count.

    a.   b.

    c.   d.

    e.

6. List and describe the three stages of formation of a blood clot.

   a.

   b.

   c.

A QUESTION TO MAKE YOU THINK

In some parts of the world, children do not get enough protein in their diets. A severe protein-deficiency type of malnutrition called *kwashiorkor* occurs. One symptom is extreme distension of the abdomen. Based upon your knowledge of plasma and of osmosis (from chapter three), explain the swelling associated with kwashiorkor.

ADDITIONAL STUDY

Read the Chapter Summary (pages 364 - 365). Write out the definitions of all the KEY TERMS (page 365).

Review the illustrations in your textbook and understand the answers to the questions associated with each one. The answers are on pages 366 - 367.

Having studied this chapter, close your book, put away your notes, and test yourself by **writing** the answers to the "CONCEPTS CHECKS" and "QUESTIONS FOR REVIEW". Writing the answers will force you to challenge yourself and give you confidence in your ability to write the answers on an exam.

Before an exam over this chapter, read the "Learning Objectives", page 344, and review any sections which seems problematic.

BLOOD
Robert W. Bauman, Jr., Ph.D.

Across

1. wbc without granular cytoplasm
5. liquid portion of blood
7. erythrocyte
9. development of blood
12. proteins on rbc
15. anticoagulant
17. leukocyte
18. smallest wbc
19. thrombocyte
20. clotting protein
21. secretes histamines
22. function of blood
26. proteins that react with antigens
28. severe blood loss
29. wbc which stain red
30. cancer of the blood
31. wandering wbc
34. undesirable clot
35. disorder of coagulation
36. enzyme of coagulation

Down

2. wbc with granular cytoplasm
3. blood antigen group
4. rbc
5. precursor of thrombin
6. measure of acidity
8. clotting
10. hormone of erythropoiesis
11. carries oxygen
13. most common globulin
14. most common wbc
16. reduction in rbc
19. collection of wbc
20. precursor to fibrin
23. protein which thickens the blood
24. squeezing between cells
25. stoppage of bleeding
27. largest wbc
28. percentage of blood cells
30. wbc
32. free floating clot
33. infection by protozoa

- Crosswords Plus

CHAPTER 13: THE CARDIOVASCULAR SYSTEM

There are few subjects in anatomy which are as familiar to students as the heart. You already know that the heart is a pump which moves blood through the body. You probably know that it has chambers, arteries, and valves and can suffer "attacks". This chapter will give you medical terms and information to supplement your layman's understanding. Now let's get to the "heart of the matter".

CONTENT MASTERY

THE HEART -- CHARACTERISTICS AND COVERINGS. Match these terms with their descriptions. Terms may be used more than once or not at all. Have the appropriate diagrams in front of you and locate the structures as you go. When you have worked through the matching, check your answers, and try to define or describe each of the terms as if they were a list of words to define. This kind of thorough study will pay off for you!

_____ 1. literally, pointed tip

_____ 2. literally, around the heart

_____ 3. outer membrane enclosing heart and major vessels

_____ 4. another term for parietal pericardium

_____ 5. two layered sac, the outer layer of the pericardium

_____ 6. another term for visceral pericardium

_____ 7. literally, upon the heart

_____ 8. tightly attached to heart surface, considered part of the heart wall

apex
epicardium
fibrous layer of parietal pericardium
pericardial cavity
pericarditis
pericardium
pericardial sac
serous layer of parietal pericardium

_____ 9. space between serous layer of the parietal pericardium and visceral pericardium

_____ 10. swelling of the pericardium causing a decline in lubrication and chest pain

_____ 11. outer fibrous layer of parietal pericardium

apex
epicardium
fibrous layer of parietal pericardium
pericardial cavity
pericarditis
pericardium
pericardial sac
serous layer of parietal pericardium

HEART WALL AND CHAMBERS. Read the sections which teach about the heart wall and the chambers within. Magnify your attention to each term by reading with your finger pointing to important words. Say that word aloud, and notice the spelling. This kind of reading is an investment of your time that will pay off in the long run. When you have read the sections, answer the following by writing "true" for each correct statement and by rewriting the false statements to make them true.

_____ 12. The wall of the heart is composed of *two* layers.

_____ 13. Blood vessels nourish *each* of the layers of the heart.

_____ 14. The layer of the heart which often contains fat deposits is the *endocardium*.

_____ 15. The *epicardium* makes up the bulk of the heart wall.

_____ 16. The inner layer of the heart wall which lines the chambers is the *endocardium*.

_____ 17. The *endocardium* is continuous with the endothelium of the major blood vessels entering the leaving the heart.

_____ 18. The atria are *superior* to the ventricles.

_____ 19. There are *three* chambers in the heart.

_____20. The atrial walls are *thicker* than the walls of the ventricles.

_____21. "Atria" means *entrance* room.

_____22. "Auricles" is another name for the *atria*.

_____23. Ridges in the atrial walls are called *interatrial* muscles.

_____24. The *interatrial septum* separates the right and left atria.

_____25. The foramen ovale in the fetus closes leaving a depression, the *fossa ovalis*.

_____26. The right atrium collects blood from *three* major vessels.

_____27. The left atrium collects blood from *four* pulmonary veins.

_____28. "Ventricle" means *strong stomach*.

_____29. Of the four chambers of the heart, the left ventricle contains the *thickest* wall.

_____30. The chamber responsible for pushing blood to the lungs is the *left* ventricle.

_____31. The chamber responsible for pushing blood through most of the body is the *left* ventricle.

_____32. Irregular folds of muscle in the inner wall of ventricles are called *trabeculae carnae*.

_____33. Portions of trabeculae carnae that attach to some heart valves are *ventricular* muscles.

_____ 34. The interventricular *sulcus* parallels the inner position of the interventricular septum.

_____ 35. From the outside of the heart, the separation of the atria from the ventricles can be determined by position of the *atrioventricular* sulcus.

HEART VALVES. "Congratulations! As student of the day, you will be honored at a brown bag luncheon at which you will be expected to give a 7 minute "mini-teach" on heart valves. There are projection slides and a pointer to use to enhance your presentation." To help you teach like a pro (short for "professor"), try this exercise. In your text, locate the illustrations of the four heart valves. Memorize their names and locations before you even begin to read this section. Get to know them better as you read. Fill in the following chart.

VALVE TYPE	VALVE NAME	STRUCTURAL CHARACTERISTICS	LOCATION	PERTINENT INFORMATION
atrioventricular (AV)	36.	37.	38.	chordae tendineae anchor cusps to papillary muscles
atrioventricular (AV)	39.	40.	41.	ditto
semilunar (SL)	42.	43.	44.	murmurs occur less frequently than with AV
semilunar (SL)	45.	46.	between left ventricle and the aorta	ditto

After your seven-minute presentation, there will be a brief question/answer session. To prepare, answer the following.

47. Why is the bicuspid valve also called the "mitral" valve?

48. How is blood prohibited from flowing backward through AV valves?

49. What happens if a valve is deteriorated for some reason and complete closure during contraction is not possible?

50. Do SL valves allow blood to flow only one direction like AV valves? If so, how?

BLOOD FLOW THROUGH THE HEART.   Using red and blue pencils, draw arrows on the diagram to show the flow of blood through the heart. When blood is oxygenated, use red arrows. When blood is low in oxygen, but high in carbon dioxide, use blue.

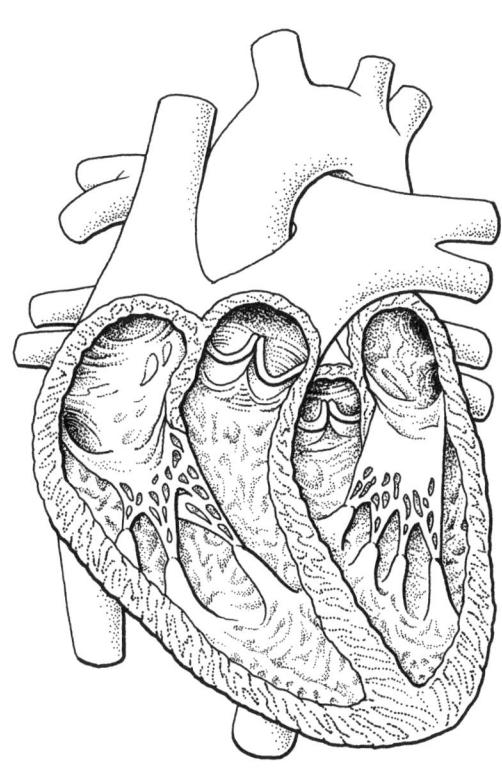

SUPPLY OF BLOOD TO THE HEART/HEART ATTACK.

_____ 51. The heart wall
   a. has a rich source of nourishment from the blood in its chambers.
   b. receives nourishment by way of diffusion of gases and nutrients from the blood in its chambers to the muscle of this active organ.
   c. has its own supply of blood vessels to meet its vital needs.
   d. none of the above are true

_____ 52. The coronary circulatory network begins at the base of the aorta where the right and left coronary arteries originate.
   a. true
   b. false

_____ 53. Most of the blood leaving the heart in coronary circulation is collected by
   a. a large vein, the coronary versicle.
   b. a large artery, the coronary artery
   c. a large vein, the coronary sinus
   d. a large artery, the arterial sinus

_____ 54. The chance of a heart attack can be lessened by reducing
   a. obesity
   b. high blood pressure
   c. cigarette smoking
   d. all of the above

_____ 55. Which is true of MI?
   a. It is a warning sign in which cardiac muscle cells are damaged
   b. Cardiac muscles are actually killed due to interrupted blood supply.
   c. As a result of MI, the heart muscle loses a degree of its strength.
   d. Both b and c are true, but a is false.

_____ 56. The most common cause of heart attack is
   a. thrombus in one of the coronary arteries
   b. an embolus
   c. a spasm in one of the coronary arteries
   d. hypertension

_____ 57. Atherosclerosis increases the chances of an MI occurring because it
   a. increases blood cholesterol
   b. narrows the coronary vessel openings and encourages the formation of clots
   c. is the replacement of dead cells with scar tissue which is less flexible than live cells and restricts blood flow
   d. reduces the oxygen supply to the coronary muscle tissue

_____58. Which of the following may be effective in preventing myocardial infarction?
   a. quitting smoking
   b. regular exercise
   c. taking low dosages of aspirin daily
   d. all of the above have been recommended

HEART PHYSIOLOGY. This section is a very clear and fascinating presentation of heart physiology. As you read it, use an effective note-taking method to help you stay focused on your reading. On a sheet of notebook paper, creased vertically, write terms and short questions on the left. Write phrases on the right that would define the term or answer the questions. When you have completed the section, complete the following matching questions (which will certainly look similar to your study notes -- good job!).

_____ 59. contraction of both atria, followed by both ventricles

_____ 60. the state of contraction

_____ 61. the state of relaxation

_____ 62. the "clicking" shut of the heart valves

_____ 63. in a "lub-dup," the "lub" indicates these (two)

_____ 64. the "dup" indicates this

_____ 65. "pacemaker"

_____ 66. once activated, generates an impulse down the AV bundle

_____ 67. AV bundle

_____ 68. carry impulses to stimulate almost simultaneous ventricular systole

_____ 69. measures electrical events during a cardiac cycle

acetylcholine
artificial pacemaker
atrioventricular (AV) node
baroreceptors
bundle of His
cardiac output
cardiac hypertrophy
cardiac cycle
closing of the AV valves
closing of the SL valves
diastole
electrocardiogram
heart rate
heart sounds
norepinephrine
P wave
Purkinje fibers
QRS complex
sinoatrial (SA) node
Starling's law of the heart
stroke volume
systole
T wave
ventricular systole

247

_____ 70. in an ECG, when the atrial myocardium is depolarized

_____ 71. in an ECG, depolarization of the ventricles

_____ 72. in an ECG, repolarization of the ventricular myocardium

_____ 73. surgically inserted mechanical device which stimulates the heart at regular frequencies

_____ 74. number of cardiac cycles per minute

_____ 75. a measurement of the volume of blood leaving the ventricles during their contraction

_____ 76. heart rate x stroke vol.

_____ 77. explanation dealing with stretching and contraction

_____ 78. thickening of ventricular walls due to years of exercise

_____ 79. slows the heart rate

_____ 80. speeds the heart rate

_____ 81. receptors in aorta and carotid arteries which detect blood pressure fluctuations

acetylcholine
artificial pacemaker
atrioventricular (AV) node
baroreceptors
bundle of His
cardiac output
cardiac hypertrophy
cardiac cycle
closing of the AV valves
closing of the SL valves
diastole
electrocardiogram
heart rate
heart sounds
norepinephrine
P wave
Purkinje fibers
QRS complex
sinoatrial (SA) node
Starling's law of the heart
stroke volume
systole
T wave
ventricular systole

ARTERIES AND ARTERIOLES. Learning the vessels of the body is like learning the names of streets on a travel route; sometimes street names change as one enters a new neighborhood. Select the best choice from those offered in parenthesis. Circle your answers with a pencil.

82. Arteries are vessels that transport blood (away from/to) the heart.

83. The thinnest vessel that can be grouped as a type of artery is an (artery/arteriole).

84. The outer layer of the walls of arteries is the (lumen/tunica adventitia).

85. The inner layer of the walls of arteries is the (tunica intima/tunica adventitia).

86. The thin layer of connective tissue that anchors the artery to neighboring structures is the (tunica media/tunica adventitia).

87. The two major properties of arteries, that is, contractility and elasticity, are accomplished largely because of the (tunica media/tunica intima).

88. Nerve fibers in the smooth muscle of the tunica media whose impulses cause a decrease in the diameter of the artery are (sympathetic fibers/vasomotor fibers).

89. When the smooth muscle relaxes, resulting in the size of the lumen returning to its resting size, this state is called (vasoconstriction/vasodilation).

90. The smallest vessels in the body are called (capillaries/arterioles).

91. The diameter of a capillary is approximately (one fingernail/one erythrocyte) wide.

CAPILLARIES. Read the four paragraphs that describe capillaries. Then answer the following like an expert!

92. Describe the size of a capillary (diameter).

93. What is the major function of capillaries?

94. What anatomical make-up of a capillary allows it to be successful in its assignment?

95. Use the term "diffusion" in reference to capillaries.

96. What is the function of the precapillary sphincter?

97. Describe capillary beds, thoroughfare channels, and true capillaries.

VENULES AND VEINS. As you continue your travel along the route of blood flowing through the vessels of the body, select the best answer in parenthesis for these statements. Circle your answer with a pencil.

98. Venules and veins are three-layered tubes that carry blood (toward/away from) the heart.

99. The smaller of the two types of veins is the (vein/venule).

100. (All/Most) veins carry deoxygenated blood.

101. Pulmonary veins carry (oxygenated/deoxygenated) blood.
102. Compared to arteries, the lumens of veins are (larger/smaller)

103. A property of veins which is not found in arteries is (elasticity/distensibility).

104. Varicose veins are those which have (weakened valves/torn muscle walls).

105. Varicose veins in the anal canal are called (venosclerosis/hemorrhoids).

106. Veins help blood return to the heart by having one-way valves. (true/false)

107. Breathing helps move blood through veins. (true/false)

108. Highly specialized veins responsible for storing blood are called venous (pumps/sinuses).

BLOOD PRESSURE. The maintenance of healthy parameters of blood pressure is critical to good health. The physiology of this primary force that pushes blood through the body is discussed fully in the next sections of text. All students should be interested in this, but those wishing to pursue nursing or medical fields will find this especially useful. As you read and study, in your imagination place yourself in a medical uniform taking someone's blood pressure. The patient asks you question after question about blood pressure. Become an expert. Be able to teach the "public" about this fascinating and practical topic! Answer the following multiple choice to help you become that expert.

_____109. If a patient has a blood pressure of 80 mm Hg, this means
   a. the pressure in his capillaries is 80 pounds per millimeter.
   b. the pressure in his major arteries is high grade (Hg)
   c. the pressure in his major arteries is great enough to lift a column of mercury a distance of 80 millimeter.
   d. the pressure in his vessels is great enough to push 80 millimeters of blood.

_____110. The pressure in veins
   a. is actually more that the pressure in arteries
   b. is actually less than the pressure in arteries

_____111. Systolic pressure is
   a. the peak of pressure (in the aorta)
   b. the lowest amount of pressure (in the aorta)
   c. the peak of pressure (in the vena cava)
   d. the lowest amount of pressure (in the vena cava)

_____112. Diastolic pressure is
   a. greater than systolic pressure
   b. less than systolic pressure
   c. a measurement of pressure during ventricular relaxation
   d. Both b and c are correct.

_____113. The apparatus used for measuring systolic and diastolic pressure is the
   a. Korotkoff
   b. Stethoscope
   c. Sphygmomanometer
   d. Both b and c are correct.

_____114. When the resting adult pulse rate exceeds 100 beats per minute, the condition it called
   a. ventricular systole
   b. tachycardia
   c. bradycardia
   d. peripheral resistance

_____115. When the resting adult pulse rate drops below 60 beats per minute, the condition is called
   a. ventricular systole
   b. tachycardia
   c. bradycardia
   d. peripheral resistance

_____116. Which of the following is true about cardiac output?
   a. An increase in cardiac output increases blood pressure.
   b. A decrease in cardiac output decreases blood pressure.
   c. Cardiac output represents the amount of blood pushed into the arterial system at a given time.
   d. All of the above are true.

_____117. Which of these is true concerning peripheral resistance?
   a. In general the smaller the diameter of the inside of the blood vessel, the smaller the peripheral resistance.
   b. Thin, less viscous blood increases peripheral resistance.
   c. Any changes that alter peripheral resistance in a blood vessel influence blood pressure.
   d. All of the above are true.

_____118. Which of the following analogies describes a situation like that of increased peripheral resistance of the blood?
   a. drinking lemonade through a large straw
   b. drinking a milk shake through a thin straw

_____119. What is an atheroma?
   a. a distinct smell accompanying a heart attack
   b. the leading cause of death in the United States
   c. the degeneration of the artery wall surface
   d. a collection of fats and cholesterol along the inner wall of arteries

_____120. Which of the following is most like atherosclerosis?
   a. the buildup of plaque on the teeth
   b. the buildup of mineral deposits on the inside of a water pipe
   c. the ballooning of a portion of tire leading to a blow-out
   d. closing the cross bar in a toll gate

_____121. Which is not a factor that affects blood pressure?
   a. blood volume
   b. cardiac output
   c. peripheral resistance
   d. percentage of erythrocytes

____122. Select the false statement concerning hormonal control of blood pressure.
   a. Epinephrine increases blood pressure by stimulating heart rate and thus increasing cardiac output.
   b. Norepinephrine increases blood pressure by increasing peripheral resistance.
   c. The atrial natriuretic factor reduces blood pressure by causing the kidneys to excrete sodium and water.
   d. Antidiuretic hormone decreases blood volume to lower blood pressure.

____123. How do the kidneys aid in maintaining stable blood pressure?
   a. respond to high blood pressure by conserving water.
   b. respond to low blood pressure by producing less urine.
   c. respond to low blood pressure by releasing renin which eventually causes more peripheral resistance.
   d. both b and c.

____124. Select statement(s) that are true concerning hypertension. Write the letters of each true statement in the blank.
   a. Hypertension is high blood pressure.
   b. Hypertension often results in reduced stress to arteries, leading to hemorrhage.
   c. As arteriosclerosis increases, hypertension also increases.
   d. A controlled diet is often recommended for persons with hypertension.

CAPILLARY EXCHANGE. The exchange of gases, food molecules, and waste happens at the capillaries. After reading this material, match the terms with their descriptions.

_____ 125. process by which materials move across capillary walls

arterial end
diffusion
hydrostatic pressure
osmotic pressure
venous end

_____ 126. a force which controls the amount of fluid that moves across capillary walls

_____ 127. pressure within capillaries that tends to push fluid into the extracellular space

_____ 128. the area of the capillaries where more fluid enters by absorption

_____ 129. the area of the capillaries where more fluid is pushed out by filtration

253

CIRCULATORY PATHWAYS. The majority of the study work for this topic will be done in the "LABEL AND LIST" section of this study guide. Other pertinent and interesting details will be picked up here to be matched as you read.

_____ 130. consists of the vessels that service all body organs except the lungs

_____ 131. consists of the vessels that service heart and lungs

portal system
aorta
systemic circulation
pulmonary circulation
great saphenous vein

_____ 132. largest artery of the body

_____ 133. longest vein in the body

_____ 134. generally, when blood flows from one capillary network to another instead of toward the heart

HOMEOSTASIS. Answer the following questions.

135. What is circulatory shock? Give its symptoms as part of your answer.

136. How does the circulatory system react to restore homeostatic balance after circulatory shock?

CLINICAL TERMS OF THE CARDIOVASCULAR SYSTEM. Match the term with its description. Terms may be used more than once or not at all.

137. heart beats with very irregular rhythm because of chaos in the heart conduction system

138. sac in vessel wall due to stretching

139. heart failure

140. leaky heart valve

141. super fast heart rate

142. causes angina pectoris

143. pain or tightness in the chest due to too little blood to the heart itself

144. infection of the pericardial sac

145. inherited narrow heart valves leading to murmurs

146. inflammation of veins

147. infection of heart muscle

148. congenital disease, opening between pulmonary artery and aorta

149. high blood pressure

150. failure of SA or AV node to generate impulses

aneurysm
angina pectoris
arteriosclerosis
atherosclerosis
bacterial endocarditis
bradycardia
cardiac arrhythmias
coarctation of the aorta
congestive heart failure
heart fibrillation
heart flutter
heart block
hypertension
ischemia
murmur
myocarditis
patent ductus arteriosus
pericarditis
phlebitis
septal defects
stenosis
tachycardia

_____ 151. congenital disease, narrowed aorta, more blood to upper parts of body than lower

_____ 152. irregularity or loss of heartbeat

_____ 153. bacterial infection of inner lining of the heart and valves

_____ 154. narrowing of arteries due to plaques along the lumen

_____ 155. loss of elasticity of the arteries

_____ 156. congenital defects resulting in "hole in the heart" in the form of incomplete closure of septum

_____ 157. may lead to a tear in a vessel or heart chamber

_____ 158. heartbeat less than 60 beats per minute

_____ 159. heartbeat greater than 100 beats per minute

aneurysm
angina pectoris
arteriosclerosis
atherosclerosis
bacterial endocarditis
bradycardia
cardiac arrhythmias
coarctation of the aorta
congestive heart failure
heart fibrillation
heart flutter
heart block
hypertension
ischemia
murmur
myocarditis
patent ductus arteriosus
pericarditis
phlebitis
septal defects
stenosis
tachycardia

LABELS AND LISTS

1. List in order the types of vessels through which an erythrocyte would travel from the aorta to the inferior vena cava. Use these words: **capillary, venule, arteriole, artery, vein**

2. List three factors that influence arterial blood pressure.

   a.

   b.

   c.

3. List the vessels through which an erythrocyte would travel from the right ventricle of the heart to the left atrium of the heart. Use the following terms: **right pulmonary artery (or left pulmonary artery), pulmonary capillaries, pulmonary veins, pulmonary trunk**

4. Label the structures.

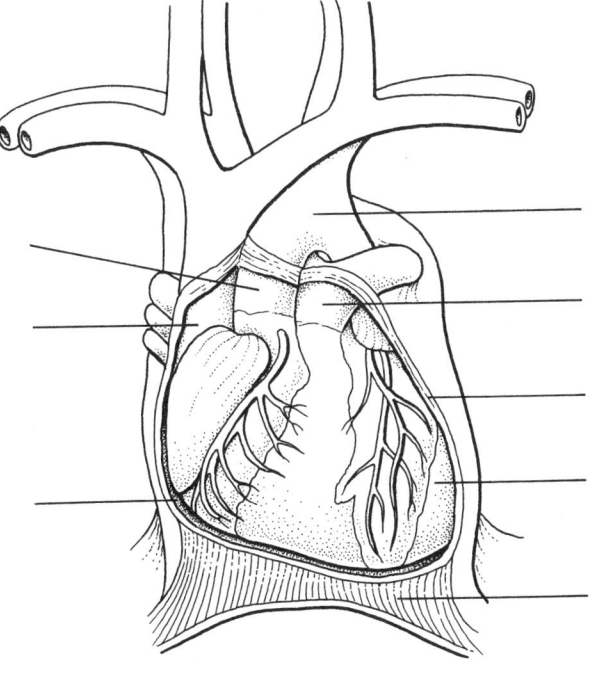

257

5. Label the internal structures of the heart.

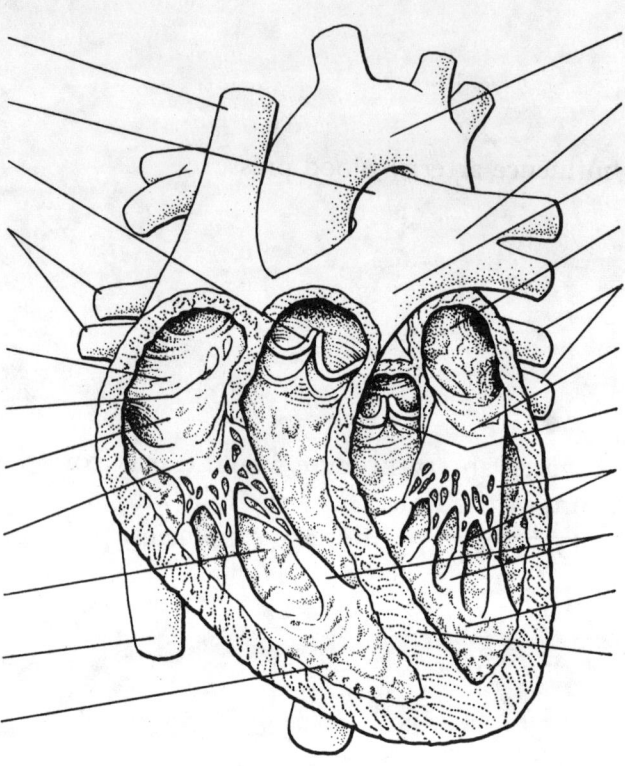

6. Label the external structures of the heart.

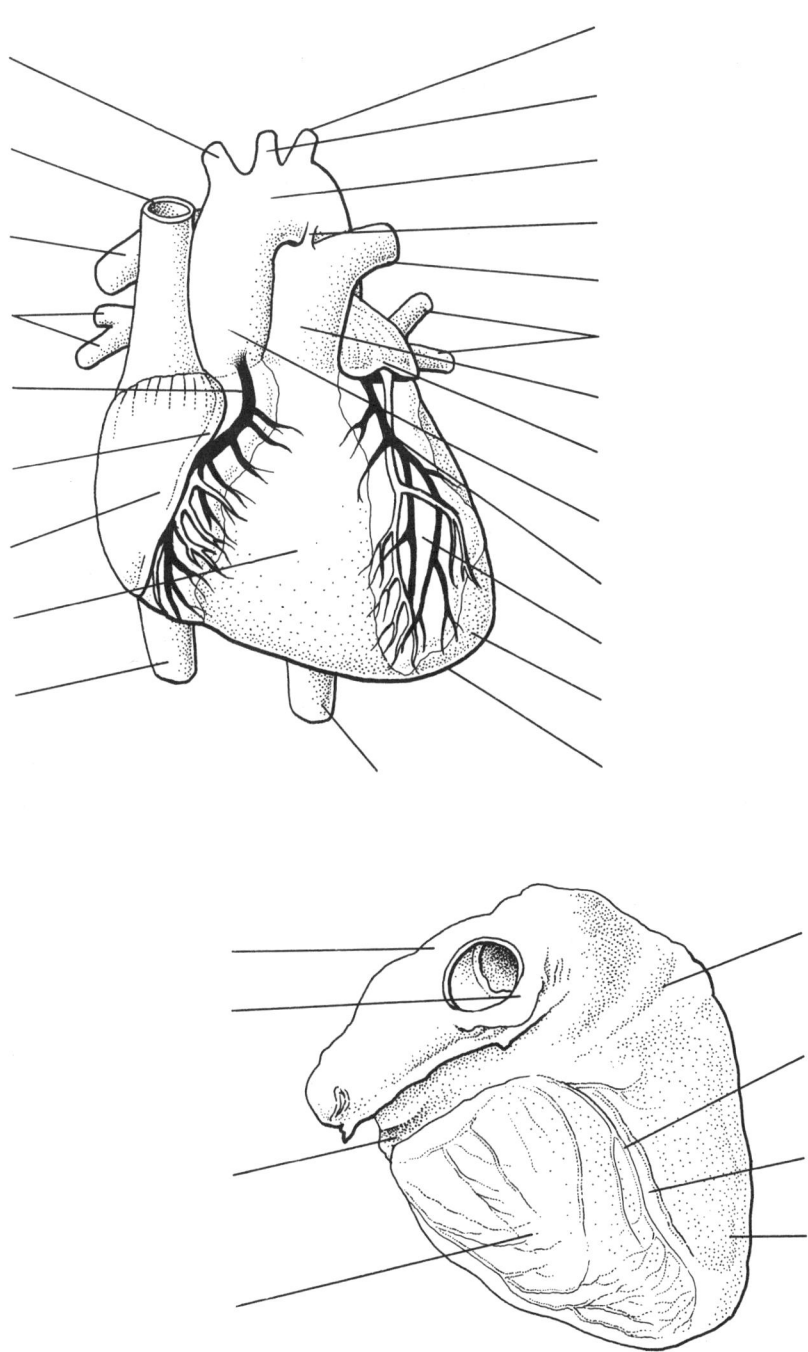

7. Sketch a normal electrocardiogram. Correctly label P, QRS, T, atrial depolarization, ventricular depolarization, and ventricular repolarization.

8. Label the aorta and its major branches.

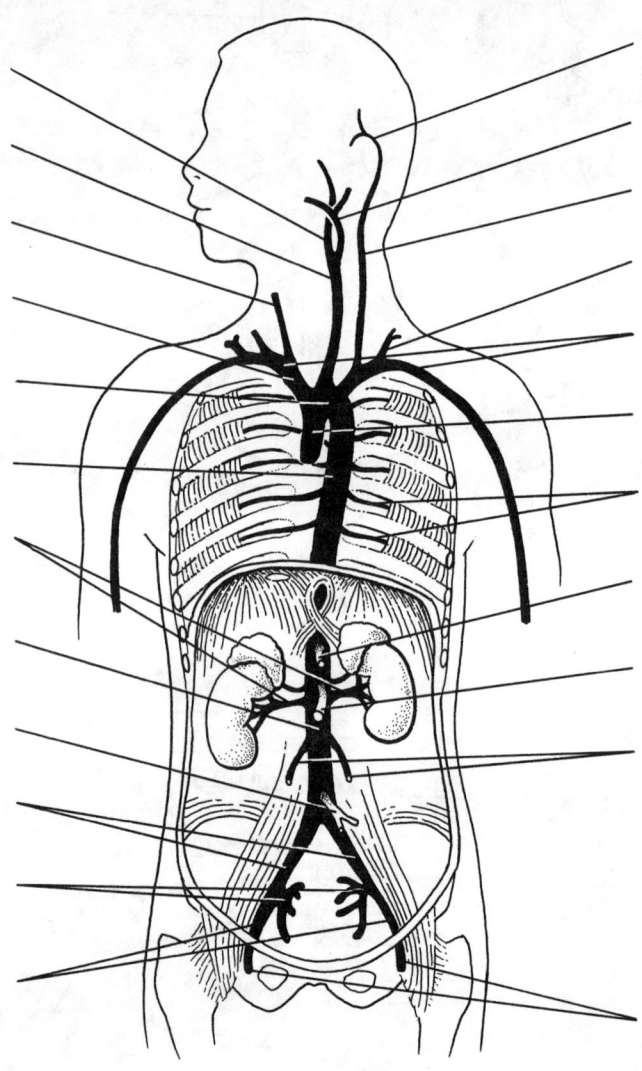

9. Label the major arteries.

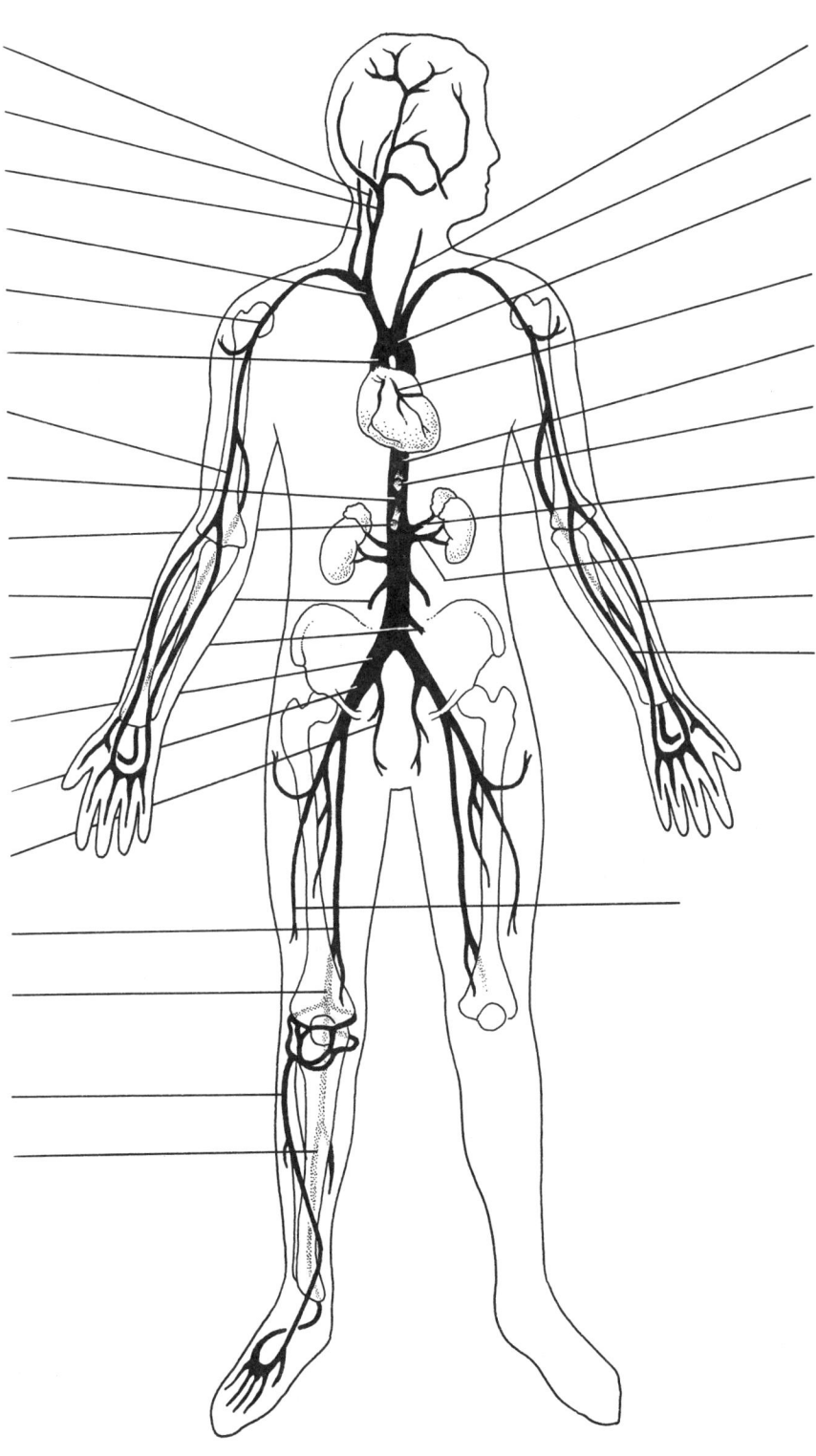

10. Label these arteries.

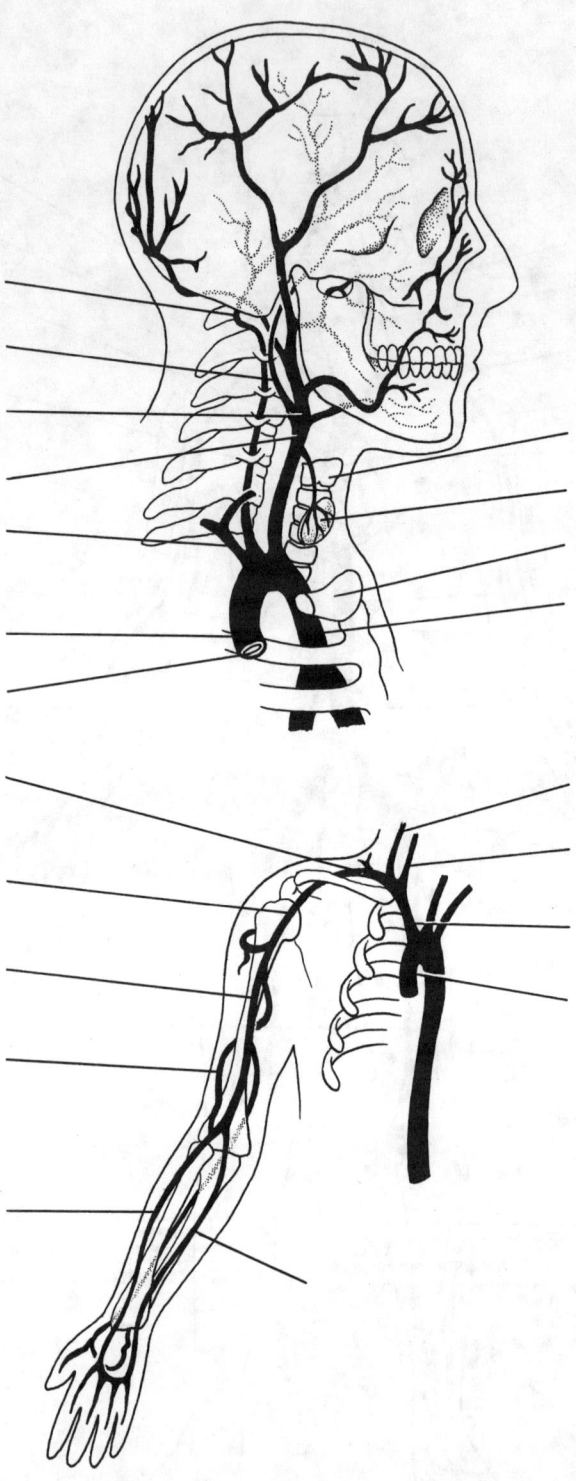

11. Label these major veins.

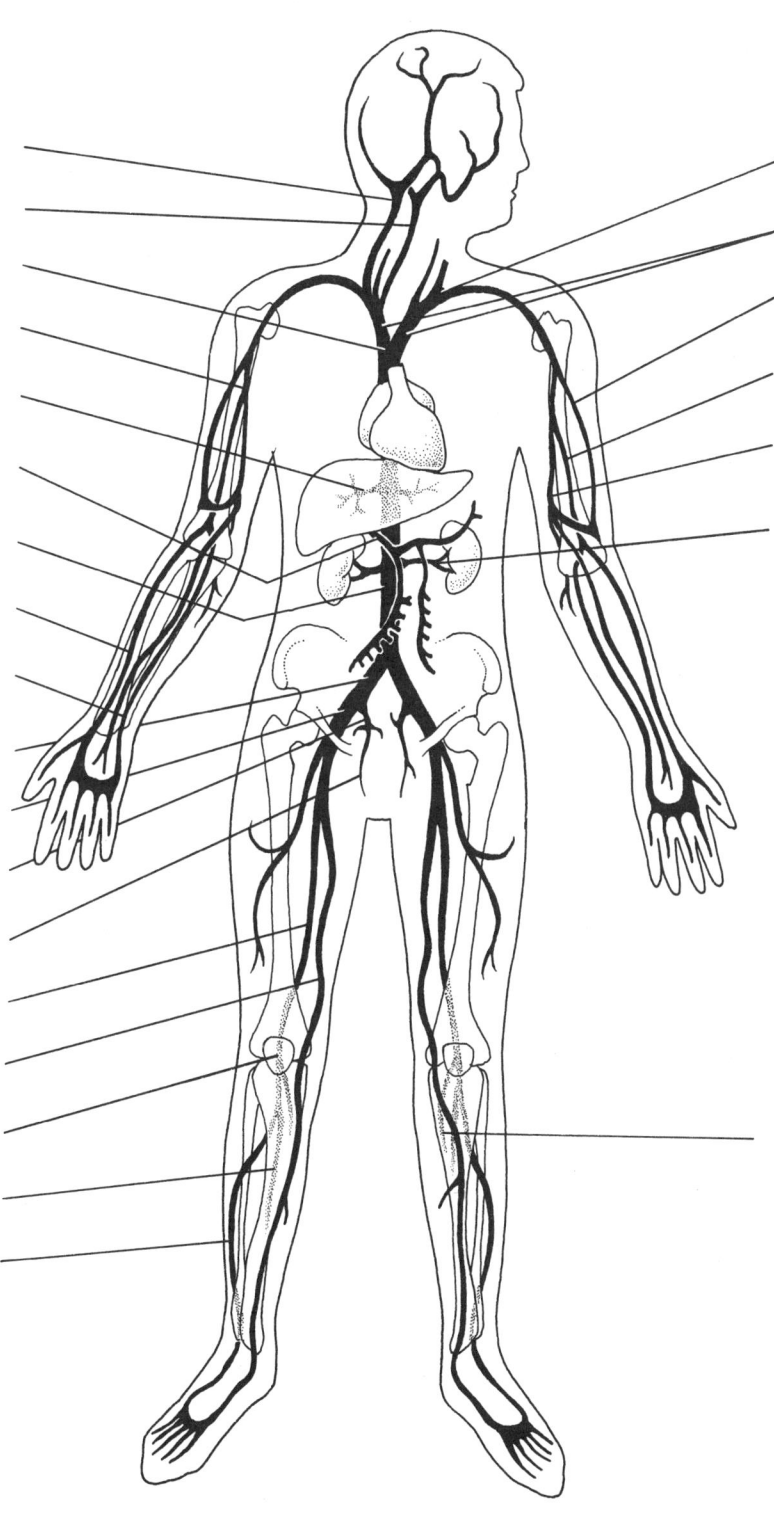

12. Label the major veins of the arm.

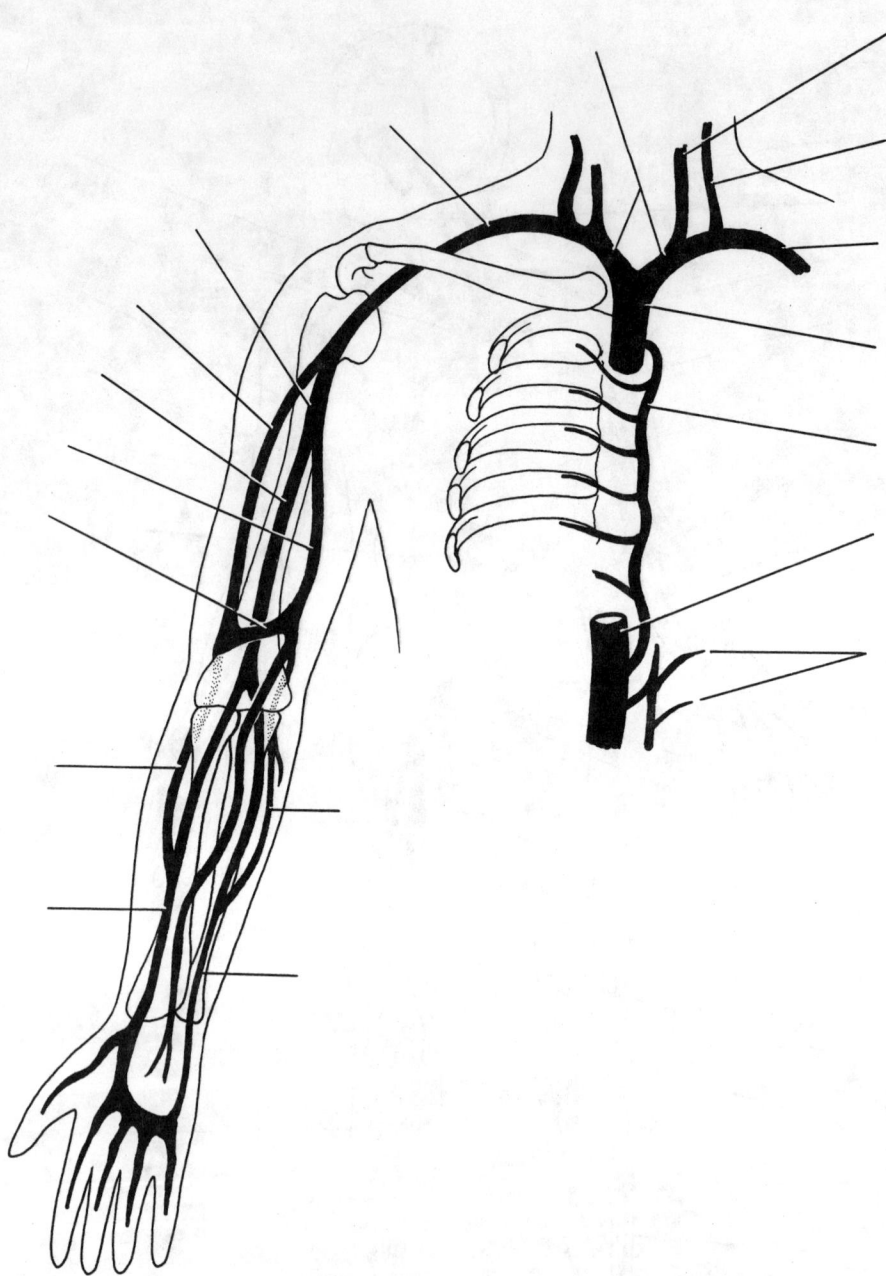

13. Label the structures of the hepatic-portal circulation.

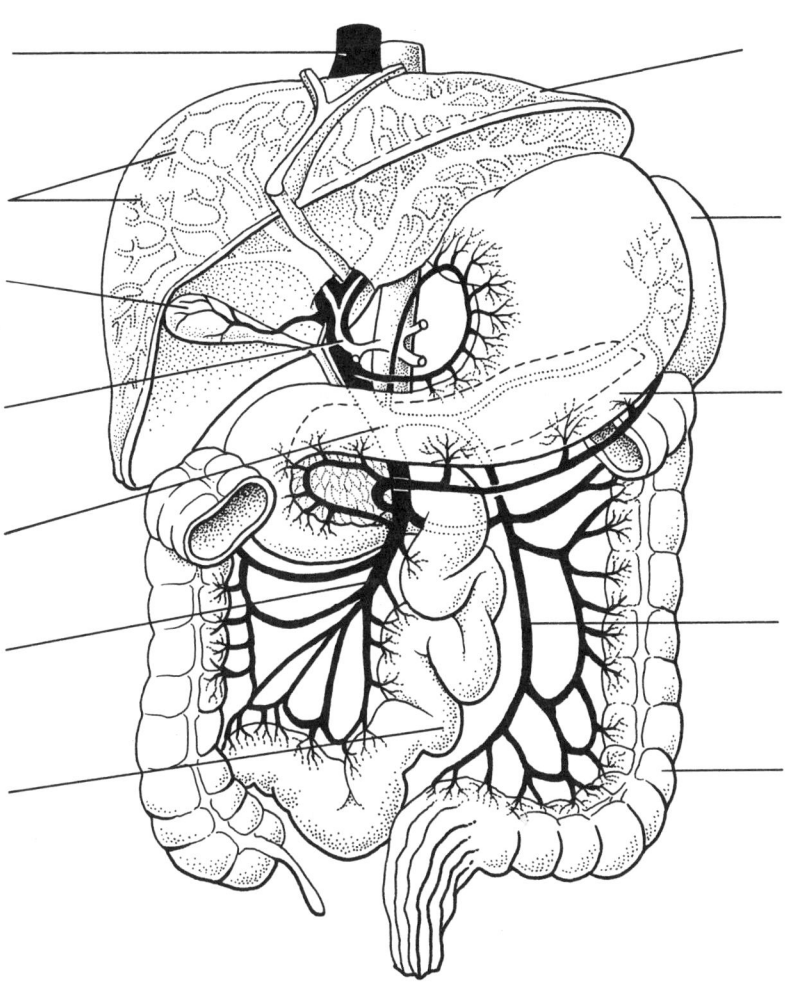

QUESTIONS TO MAKE YOU THINK

1. A student asks, "The P wave and the QRS complex of an electrocardiogram represent atrial and ventricular depolarization respectively. The T wave represents ventricular repolarization. Does atrial repolarization occur? If so, where is its wave?" Pretend that you are the teacher, and correctly answer the student's question.

2. What are the differences between skeletal muscle tissue, smooth muscle tissue, and cardiac muscle tissue?

3. Two students, Vomer and Maxilla, have a discussion concerning electrocardiograms and heart sounds. Vomer states that the P wave, the lub sound, and atrial systole are the same thing. Maxilla thinks Vomer is incorrect, but cannot explain her reasons. Write an explanation that will help Maxilla correct Vomer's statement.

4. How are the muscle fibers of the AV bundle similar to nerve fibers? How are they different?

5. Contraction of skeletal muscle cells is initiated by depolarization of the cell membrane due to the release of a neurotransmitter from a neuron. On the other hand, "Cardiac muscle cells have an inherent ability to contract." (page 379) How can the cardiac muscle cells of the SA node depolarize and contract even without a nervous supply?

6. Based upon your knowledge of skeletal muscle contraction (chapter eight), what role do the ions of sodium, potassium, and calcium likely play in contraction of the heart?

7. "Between adjacent cardiac cells the plasma membrane is thickened, forming a special junction called an intercalated disk." (Chapter 4, page 100) Now that you have studied the physiology of the heart, what function do you think is played by the intercalated disks?

ADDITIONAL STUDY

Read the Chapter Summary (pages 406 - 408). Write out the definitions of all the KEY TERMS (page 408).

Review the illustrations in your textbook and understand the answers to the questions associated with each one. The answers are on pages 410 - 411.

Having studied this chapter, close your book, put away your notes, and test yourself by **writing** the answers to the "CONCEPTS CHECKS" and "QUESTIONS FOR REVIEW" in your text. Writing the answers will force you to challenge yourself.

The day before an exam over this chapter, read the "Learning Objectives", page 368, and review any of the sections which you think will cause you a problem.

CHAPTER 14: THE LYMPHATIC SYSTEM

It is the lymphatic system that is involved with the condition known to dieters as "water retention". The lymphatic system also plays the major role in disease control.

CONTENT MASTERY

THE LYMPHATIC NETWORK. As you will see, the lymphatic network transports interstitial fluid/lymph between the cells of the body and the circulatory system. As you read this section in your text, note what you consider to be the most essential points. Review your notes before you try to answer these questions (without assistance the first time you try each question). Make it a personal challenge to get the most out of your reading!

_____1. When blood surges through capillaries, fluid is pushed through the walls carrying gases and nutrients to the body cells. Most of this interstitial fluid passes back into the blood stream. What happens to the interstitial fluid that does not diffuse back into the capillaries of the blood stream?
   a. It remains in the tissues until the blood is thinner.
   b. It is excreted by the kidneys.
   c. About 90% of the interstitial fluid moves into the lymphatic network.
   d. It is pushed into the lymphatic network and is slowly returned to the heart.

_____2. The lymphatic network begins with microscopic tubes known as
   a. lymph vessels
   b. lymphatic capillaries
   c. protein filaments
   d. lymphatic ducts

_____3. Where are lymphatic capillaries found?
   a. among capillary beds
   b. in the brain
   c. in the spinal cord
   d. in bone tissue

_____4. What prevents lymph from leaking back into the extracellular spaces?
   a. valves
   b. overlapping endothelial cells
   c. low pressure in the capillaries
   d. gaps between the endothelial cells

_____5. Which of the following is most like a lymphatic vessel in terms of structure?
   a. capillaries
   b. veins
   c. venules
   d. collecting ducts

_____6. Which of the following is NOT true of lymph nodes?
   a. They gradually increase in size and eventually merge into collecting ducts.
   b. They are small.
   c. They are oval in shape.
   d. They receive and pass on lymph by way of lymphatic vessels.

_____7. Numerous lymphatic vessels merge to form
   a. lymphatic capillaries.
   b. lymphatic nodes.
   c. collecting ducts.
   d. lymphatic trunks.

_____8. The main collecting vessel for the lymphatic network, draining lymph from the left side of the body is the
   a. thoracic duct
   b. right lymphatic duct
   c. lymphatic duct
   d. cranial duct

_____9. Which duct empties lymph into the left subclavian vein?
   a. thoracic duct
   b. right lymphatic duct

_____10. Which duct empties lymph into the right subclavian vein?
   a. thoracic duct
   b. right lymphatic duct

_____11. Which of the following is FALSE concerning the movement of lymph through the body?
   a. Pressure gradients are essential in the movement of lymph.
   b. The accumulation of protein in the interstitial fluid affects lymph movement.
   c. Lifting weights affects lymph movement.
   d. Blood pressure is a major factor in the movement of lymph.

OTHER LYMPHATIC ORGANS. These organs are the "unsung heroes" of the body which mount an immune response against disease-causing microorganisms.

_____ 12. special connective tissue in all lymph organs	afferent lymphatic vessels
	efferent lymphatic vessels
	fibrous capsule
_____ 13. type of white blood cell in all lymph organs	hilus
	lingual tonsils
	lymphoid tissue
_____ 14. vessels that enter lymph nodes	lymphocyte
	lymph nodules
	lymph nodes
_____ 15. the concave margin of a lymph node	palatine tonsils
	Peyer's patches
	pharyngeal tonsils
_____ 16. vessels that leave lymph node	red pulp
	spleen
	thymus
_____ 17. "shell" around lymph node	white pulp

_____ 18. clusters of lymphocytes within a lymph node

_____ 19. the largest of the lymphatic organs

_____ 20. area of the spleen filled with red blood cells

_____ 21. area of spleen containing white blood cells

_____ 22. functions in removing defective red blood cells from the blood

_____ 23. in children, this gland promotes the maturity of lymphocytes

_____ 24. lymphatic organs located on the back end of the palate in the mouth

_____ 25. lymphatic organs located in the nasopharynx

_____ 26. lymphatic organs located at the base of the tongue

_____ 27. clusters of lymphoid tissue in the distal end of the wall of the small intestine

afferent lymphatic vessels
efferent lymphatic vessels
fibrous capsule
hilus
lingual tonsils
lymphoid tissue
lymphocyte
lymph nodules
lymph nodes
palatine tonsils
Peyer's patches
pharyngeal tonsils
red pulp
spleen
thymus
white pulp

THE DEFENSE MECHANISMS OF THE BODY. The study of immunology is enough to make one feel total awe at the life-saving immune system of this incredible machine we live in. As you study the text, fill in the blanks with the appropriate vocabulary words.

_____ 28. In immunology, cells that belong to the body are called ...

_____ 29. In immunology, cells that do not belong to the body are called ...

_____ 30. ... proteins on the plasma membrane of cells that are coded by MHC genes.

_____ 31. The type of body defense that helps to prevent the entrance of foreign materials into the body is called ...

_____ 32. The production of antibodies is an example of a kind of defense mechanism called a ...

_____ 33. The most important physical barrier that prevents foreign substances from entering the body is the ...

_____ 34. Another important physical barrier that prevents foreign substances from penetrating the body is the ... membrane which traps microorganisms.

_____ 35. Generally speaking, when a disease-causing microorganism makes it through the body's defenses, the result is a ...

_____ 36. ... is the ingestion and destruction of particles by specialized cells.

_____ 37. The most active phagocytes are monocytes and ...

_____ 38. A mature form of a monocyte is a ...

_____ 39. ... fix themselves to the walls of blood vessels (sort of "lying in wait" for invaders).

_____ 40. The cleansing of the blood by macrophages constitute an important part of the ...

_____ 41. NK, or ... cells are a type of white blood cell that kills invading foreign cells nonspecifically by a method other than phagocytosis.

_____ 42. Two groups of normal proteins that aide in nonspecific defense are ... and interferons.

_____ 43. The body's main defense against viruses is a groups of proteins called ...

_____ 44. ... aids the defense process by preventing the spread of disease agents, removing dead cells, and assisting in tissue repair.

_____ 45. The leaking of fluids into the extracellular space is known as ...

_____ 46. Heat is produced during the inflammatory response because of increased blood flow from warmer, deeper areas of the body. (true/false)

Now that you have thoroughly studied this material, answer the following "short answer" questions as if you were the teacher answering a student's inquiry.

47. In immunological terms, what is complement, and how does it help the body?

48. How does an interferon work?

49. Why does redness occur when a bowling ball falls on a big toe?

50. Why would a swelling occur after someone is hit with a baseball?

51. Why is vasodilation at an injury site useful in the healing process?

TIP FOR THE DAY. May we suggest a study session with the table in your text which describes the "Components of Nonspecific Mechanisms"? Be able to list the eight components. Then, covering the descriptions and functions, write those answers for each component. Think of how many questions you can answer with this handy, well-organized chunk of learning!

SPECIFIC MECHANISMS: THE IMMUNE RESPONSE. As you read the details of the immune response, use this reading pacer to assist you in focusing your attention. Underline the answer of your choice given in parenthesis.

52. The immune response relies upon the ability of (leukocytes/lymphocytes) to recognize specific (antigens/antibodies).

53. Immunity is a (specific/nonspecific) defense mechanism.

54. The type of immunity in which cells provide the main defensive strategy is called (cell-mediated/humoral) immunity.

55. (Antigens/Antibodies) are chemicals that cause a response when they enter the body.

56. When the immune system destroys the body's healthy cells, this problem is called (tissue rejection/autoimmune disease).

57. Antibodies belong to a family of proteins known as (immunoglobulins/ antigen-antibody complex).

58. As the body encounters new forms of antigens, new type of antibodies to bind with the antigens are made. (true/false)

59. The immunoglobulin that is involved in the allergic response is (IgG/IgE).

60. The immunoglobulin found in tears, saliva, and mucus is (IgA/IgM).

61. The immunoglobulin found in the intestines is (IgG/IgD).

62. A T cell is a (macrophage/lymphocyte).

STUDY HINT: The five classes of immunoglobulins can be recalled by the use of the acronym "GAMED" - IgG, IgA, IgM, IgE, IgD.

63. Fill in the following chart:

COMPONENT	DESCRIPTION	FUNCTION
Monocyte		
T cell		
Killer T cell		
Helper T cell		
Suppressor T cell		
Memory T cell		
B cell		
Plasma cell		
Memory B cell		
Antibody		

CELL-MEDIATED IMMUNITY. Figure 14-10 is your visual aid as you explain cell-mediated immunity to an imaginary group of students. Read the paragraphs from the text aloud and use your finger as a pointer to direct their attention. "Bluff" your way through it the first time. Then try it again from memory. This is excellent preparation for an exam!

HUMORAL IMMUNITY. Again, play the part of the "great imposter". You have foolishly bragged to your date that you are a third-year medical student. The result is an invitation to speak to a class of physiology students. Provided with only an overhead projection slide of Figure 14-11 and your textbook, talk your way through humoral immunity by reading the material with authority while you point to pertinent parts of the figure. Answer the following student questions about both cell-mediated and humoral immunity.

_____ 64. What chemical (usually a large molecule such as a protein) triggers a specific immune response?

_____ 65. Which cells phagocytize an antigen?

_____ 66. Which cells process and display antigen?

_____ 67. Which cells are sensitized by presentation of an antigen on a macrophage membrane? (two)

_____ 68. Which T cells are capable of killing specific non-self invaders?

_____ 69. Which cells release lymphotoxins?

_____ 70. Which T cells assist killer T cells and also increase humoral immunity?

_____ 71. Which cells serve to rapidly instigate cell-mediated immunity upon subsequent invasions by the same pathogen?

_____ 72. Which cells inhibit helper T cells?

_____ 73. Which cells provide the body with humoral immunity? (two possible answers)

_____ 74. Which cells sensitizes the B cells?

_____ 75. Which cells secrete antibodies?

antibody
antigen
B cells
complement
helper T cells
killer T cells
macrophages
memory B cells
memory T cells
monocytes
natural killer cells
plasma cells
suppressor T cells
T cells

_____ 76. Which cells act to rapidly initiate humoral immunity upon subsequent exposure to an antigen?

_____ 77. Which protein molecules in the blood are capable of destroying cells which have been "flagged" by antibody?

_____ 78. What chemical is released by plasma cells?

_____ 79. What cells kill antigen-carrying cells without phagocytosis?

_____ 80. What two chemicals destroy bacteria and other cells?

_____ 81. Which cells produce plasma cells?

_____ 82. Which cells are the first to identify antigen as non-self?

_____ 83. Which cells are immature macrophages?

_____ 84. In general, which cells are also known as lymphocytes? (two)

antibody
antigen
B cells
complement
helper T cells
killer T cells
macrophages
memory B cells
memory T cells
monocytes
natural killer cells
plasma cells
suppressor T cells
T cells

AUTOIMMUNE DISEASES.

_____ 85. Another term for autoimmune disease is
   a. AIDS
   b. immunologic tolerance
   c. autoimmunity
   d. self destruction

_____ 86. What results when some leukocytes lose the ability to detect MHC proteins?
   a. They lose their immunologic tolerance.
   b. Immunologic mechanisms are activated against self molecules.
   c. Suppressor T cells are produced in exaggerated quantities.
   d. Both a and b above.

_____87. What is the cause of autoimmune diseases?
   a. The cause is probably genetic.
   b. The cause is probably viral.
   c. Poor diet causes the immune response to go awry.
   d. Scientists do not have a clue as to the cause.

_____88. All autoimmune diseases are terminal.
   a. true
   b. false

_____89. Autoimmune diseases generally last for a lifetime.
   a. true
   b. false

_____90. The allergic response involves the release of antibodies by plasma cells.
   a. true
   b. false

_____91. An allergen is
   a. an antibody formed by plasma cells in persons who are allergic.
   b. an antigen which causes a body reaction in persons who are allergic.
   c. an allergic response which provides a tolerance to most nonself molecules.
   d. a person who has an allergic response to an antigen.

_____92. The antibody which is the "culprit" for people who are allergic is
   a. IgG
   b. IgM
   c. IgE
   d. IgA

_____93. Which of the following combinations results in the release of inflammatory histamines, serotonins, and prostaglandins in allergic responses?
   a. mast cells + IgE
   b. mucous membranes + IgG
   c. urticaria + antihistamines
   d. serotonin + skin

_____94. Hay fever is also termed
   a. allergic conjunctivitis
   b. asthma
   c. anaphylaxis
   d. allergic rhinitis

_____ 95. Watery eyes are also known as
   a. allergic conjunctivitis
   b. asthma
   c. anaphylaxis
   d. allergic rhinitis

ACQUIRED IMMUNITY. Fill in the following chart which will summarize the four types of acquired immunity.

TYPE OF IMMUNITY	DESCRIPTION	EXAMPLES
naturally acquired active immunity	96.	97.
naturally acquired passive immunity	98.	99.
artificially acquired active immunity	100.	101.
artificially acquired passive immunity	102.	103.

HINT: Active immunity (of both types) involves the *production* of antibodies by the body. Passive immunity (of both types) involves the *introduction* of preformed antibodies. *Artificial* immunity (of both types) usually involves a needle!

Match these terms with their descriptions. Terms may be used more than once or not at all.

artificially acquired active immunity
artificially acquired passive immunity
immune serum globulins
naturally acquired active immunity

naturally acquired passive immunity
vaccination
vaccine

_____ 104. develops from exposure to other people who have the disease

_____ 105. when a baby is immune to the diseases for which the mother has immunity

_____ 106. an antigen which is artificially introduced

_____ 107. an injection of antigens

_____ 108. type of immunity against chicken pox

_____ 109. antisera

_____ 110. short-term immunity by injection with foreign antibodies

HOMEOSTASIS. Maintaining homeostasis is essential to life. Homeostasis of the lymphatic system is no exception. The two functions of recycling fluids and immunity must be maintained. Read the section in your text on homeostasis and then mark the following statements as "true" or "false". Rewrite the false statements to make them true by changing the italicized words.

_____111. White bloods *cannot* be infected by disease-causing microorganisms.

_____112. Impairment of the body's ability to defend against pathogens is termed *immunodeficiency*.

_____113. Severe combined immunodeficiency, *like AIDS, can be acquired*.

_____114. Kaposi's sarcoma is a skin cancer normally seen in *young* people.

_____115. A person infected with HIV can be tested for its presence about *six* months after infection.

_____116. A person who has HIV, but no AIDS symptoms, *is not* considered infectious.

_____117. ARC, AIDS related complex, *is* the same thing as AIDS.

_____118. Usually, AIDS will be contracted within *six months* from the time of infection.

_____119. Currently, there *is no* cure for AIDs.

_____120. A effective vaccine for AIDS *has been* demonstrated.

_____121. *All* persons are capable of developing AIDS if they are exposed to contaminated body fluids.

_____122. HIV prefers to infect *macrophages*.

_____123. HIV can infect helper *T cells and monocytes*.

_____124. The typical normal symptoms of initial HIV infection include *Kaposi's sarcoma* and Pneumocystis *pneumonia*.

_____ 125. Pathogenic organisms which are normally destroyed by a healthy immune system are termed *opportunistic pathogens*.

_____ 126. HIV is transmissible through semen, *but not* through infected vaginal secretions.

_____ 127. HIV is a *strong* virus.

_____ 128. AIDS *can be* prevented.

CLINICAL TERMS OF THE LYMPHATIC SYSTEM. Match these terms with their descriptions. Terms may be used only once. Check your spelling.

_____ 129. use of tumor-specific antigens from transplanted tumors to destroy cancers

_____ 130. malignant cancer of lymphoid tissue

_____ 131. decreased neutrophil count

_____ 132. destruction of immune response

_____ 133. severe response to bee sting

_____ 134. ability of the immune response to reactivate quickly during repeated exposures to antigen

_____ 135. a child born without thymus and parathyroid glands, lacks control of infection

acquired immuno-
    deficiency diseases
anaphylaxis
atopic diseases
autoimmunity
DiGeorge's syndrome
immunization
immunosuppression
immunotherapy
Hodgkin's disease
lymphoma
monoclonal antibodies
neutropenia
splenomegaly
vaccine

283

_____ 136. cancer in lymphoid tissue occurring most often in young adults and adults over 50

_____ 137. pure antibody preparation that combines with only one type of antigen

_____ 138. given to immunize against smallpox, typhoid, polio, etc.

_____ 139. enlargement of the spleen following infectious disease

_____ 140. rheumatoid arthritis is an example

_____ 141. bronchial asthma and hay fever, for example

acquired immuno-deficiency diseases
anaphylaxis
atopic diseases
autoimmunity
DiGeorge's syndrome
immunization
immunosuppression
immunotherapy
Hodgkin's disease
lymphoma
monoclonal antibodies
neutropenia
splenomegaly
vaccine

LABEL AND LIST

1. List 5 specialized organs of the lymphatic system.

    a.   b.

    c.   d.

    e.

2. List the names and number of the tonsils.

    a.   b.

    c.

3. When histamine and serotonin are produced by damaged cells, what two main responses can be expected in the local tissues

    a.

    b.

284

4. What four symptoms characterize the inflammatory response?

   a.   b.

   c.   d.

5. List the five classes of antibodies (immunoglobulins).

   a.   b.

   c.   d.

   e.

6. List four types of T cells.

   a.   b.

   c.   d.

7. List two types of B cells.

   a.   b.

8. List four types of acquired immunity.

   a.   b.

   c.   d.

9. List four classes of people who are more susceptible to AIDS.

   a.   b.

   c.   d.

10. List six diseases which are presently used to diagnose AIDS.

    a.   b.

    c.   d.

    e.   f.

11. Label the lymphatic system.

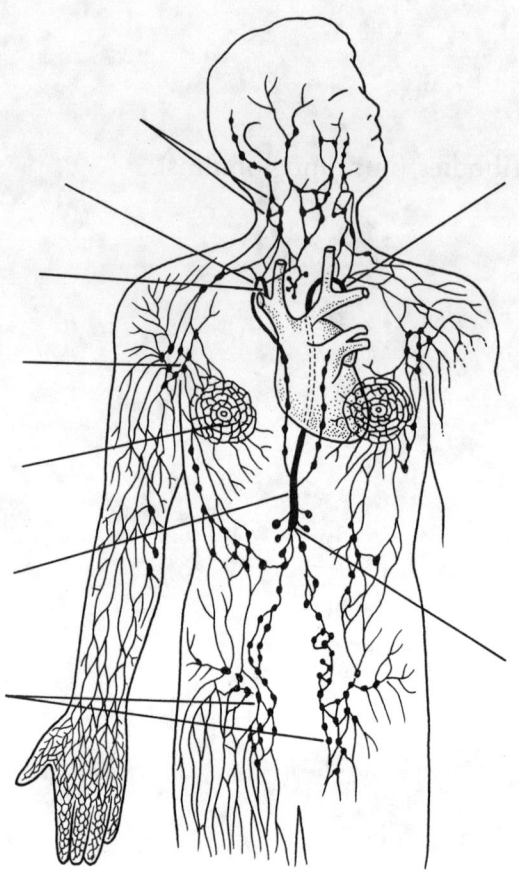

12. Label the lymph node.

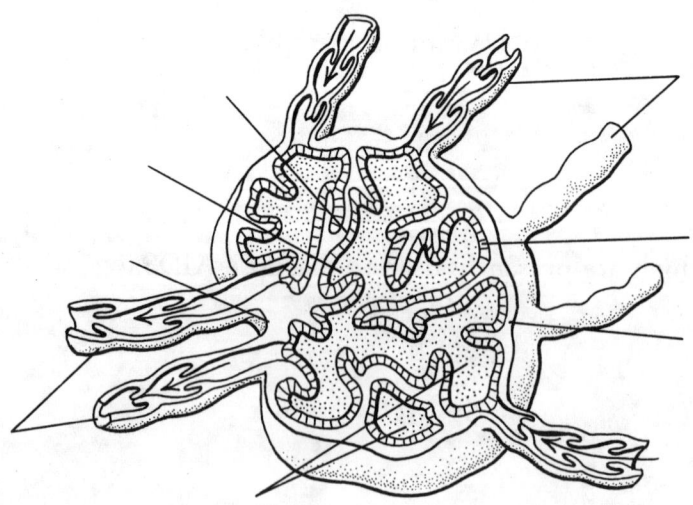

ADDITIONAL STUDY

Read the Chapter Summary (pages 436 - 437). Write out the definitions of all the KEY TERMS (page 437).

Review the illustrations in your textbook and understand the answers to the questions associated with each one. The answers are on page 439.

Having studied this chapter, close your book, put away your notes, and test yourself by **writing** the answers to the "CONCEPTS CHECKS" and "QUESTIONS FOR REVIEW" in your text. Writing the answers will force you to challenge yourself. If you can write the answers for yourself, you can probably write the answers for your professor.

The day before an exam over this chapter, read the "Learning Objectives", page 412, and review any of the sections which you think will cause you a problem.

THE LYMPHATIC SYSTEM
by Robert W. Bauman, Jr., Ph.D.

Dec 19, 1993 - Crosswords Plus

Across

2. immunodeficiency syndrome
5. proteins which prevent further viral infection of cells
8. immunity involving antibodies
10. secretory immunoglobulin
13. largest lymphatic organ
14. mass of lymphatic tissue
15. toxic chemical released by T cells
16. genes which determine 'self'
18. T cell which stops immune response
20. defense mechanism which defends against all types of pathogens
22. collection of wbc and fluid
23. T cell which produces lymphotoxin
24. immunoglobulin involved in allergies
25. expanded, inferior portion of thoracic duct
26. primary source of nonspecific immunity
28. type of immunity produced by vaccines
29. hives
32. right lymphatic ... empties into the right subclavian vein
34. fluid within lymphatics
35. substance which provokes an immune response
37. cell which is 'presensitized' to antigen
38. cell which produces antibody
40. patches of lymphoid tissue in small intestine
41. protein produced in response to antigen
42. immunity produced by receipt of preformed antibody
43. antigen which triggers allergic response
44. wbc which nonspecifically kill nonself, infected, and cancer cells by lysis

Down

1. lymphatic tissue located in the mouth and throat
3. immunoglobulin which traverses placenta
4. disease causing agents
6. endocrine gland involved in T cell maturation
7. lymphatic duct which empties the upper, right portion of the body
8. T cell which stimulate humoral immunity
9. cell which releases histamine
10. invasion of pathogens
11. most serious allergic response
12. leaking of fluid into intercellular spaces
15. chemical produced by lymphocyte which affects other wbc
17. plasma proteins which are involved in nonspecific immunity
19. redness, swelling, heat, pain
21. ingestion by cells
24. study of immunity
27. large phagocyte
30. immunity against 'self'
31. malignant tumor of lymphoid tissue
33. lymphatic duct which drains the left and lower portions of the body
35. lymphatic vessel draining into a node
36. lymphatic vessel draining a node
39. allergen lodged in an airway may trigger ...

CHAPTER 15: THE RESPIRATORY SYSTEM

You have now come to another system which is familiar to you. Learning the anatomical and physiological terms for "breathing" etc. should pose no problems!

CONTENT MASTERY

_____ 1. The movement of air between the external environment and the air sacs of the lungs is
   a. pulmonary ventilation
   b. external respiration
   c. internal respiration
   d. cellular respiration

_____ 2. The diffusion of molecules of gases between the air sacs and their capillaries is
   a. pulmonary ventilation
   b. external respiration
   c. internal respiration
   d. cellular respiration

_____ 3. The movement of oxygen and carbon dioxide between the bloodstream and body cells is called
   a. pulmonary ventilation
   b. external respiration
   c. internal respiration
   d. cellular respiration

_____ 4. Which is the correct sequence of events in respiration?
   a. internal respiration, external respiration, pulmonary ventilation
   b. external respiration, internal respiration, pulmonary ventilation
   c. pulmonary ventilation, internal respiration, external respiration
   d. pulmonary ventilation, external respiration, internal respiration

ORGANS OF THE RESPIRATORY SYSTEM. Read this introductory section in your textbook. It will help you organize a framework for more intense study in the sections to come. Answer the following.

5. When the respiratory system is considered anatomically (according to its structure), it can be divided into that two portions?

6. When the respiratory system is considered physiologically (according to its functions), it can be divided into what two zones?

NOSE. Match these terms with their descriptions. Though it may be tedious to write them, it will assist you in memory and with spelling. Resist the temptation to say to yourself, "I know that one, I don't need to write it down." Such terms and concepts tend to "disappear" from memory when you are faced with an exam question! WRITE!

_____	7. initial receiving chamber for inhaled air
_____	8. two openings at the base of the nose
_____	9. the partition which divides the nasal cavity in half
_____	10. literally, entrance chamber
_____	11. thin bones in the nasal cavity increasing its surface area
_____	12. literally, opening or passage
_____	13. blockage of ducts and sinuses causing pressure and pain

external nares
meati
nasal conchae
nasal septum
nose
sinusitis
vestibule

PHARYNX. Read about your throat. Pay attention to the illustrations, and be sure to study them in the LABEL AND LIST section of this chapter study guide. Fill in the blanks with the correctly spelled terms as you read about the pharynx.

_____ 14. The throat is also called the ...

_____ 15. The walls of the throat are formed by ...

_____ 16. The throat is lined with ... membrane.

_____ 17. Air in the nasal cavity enters the throat via the ...

_____ 18. The part of the pharynx that has openings from the eustachian tubes is the ...

_____ 19. The part of the pharynx visible in the mirror is the...

_____ 20. The part of the pharynx that touches the larynx in the neck is the...

_____ 21. Food travels over the ... and the ... portions of the pharynx.

LARYNX. Match these terms with their descriptions. Terms may be used more than once or not at all, and they must be spelled correctly.

_____ 22. the voice box

_____ 23. connects the throat to the windpipe

_____ 24. houses the vocal cords

_____ 25. literally, upper windpipe

_____ 26. literally, shield-like

_____ 27. literally, ring-like

_____ 28. literally, upon the tongue

_____ 29. "Adam's apple" in males

_____ 30. the opening into the larynx

_____ 31. cartilage which closes the glottis

_____ 32. folds in the larynx that function in swallowing but not in sound production

_____ 33. lower pair of folds in the larynx that generate sound waves by vibrating

cricoid
epiglottis
false vocal cords
glottis
larynx
thyroid
thyroid cartilage
true vocal cords

TRACHEA. Select the best answer from those suggested. Place the letter of your answer in the blanks.

_____34. The windpipe is also known as the
   a. pharynx
   b. larynx
   c. trachea
   d. vocal cord

_____35. The trachea is a tubular air passageway. Approximately how long is it?
   a. 1 inch
   b. 3 inches
   c. 12 centimeters
   d. 1.5 centimeters

_____36. Of what advantage is it that each ring of cartilage on the trachea is shaped like an open "c"?
   a. This allows for inhalation of large amounts of air in an emergency.
   b. It gives the heart ease in pumping.
   c. It allows the esophagus to expand with food.
   d. It allows for contraction of the trachea during coughing.

_____37. Which is true of the mucociliary transport system?
   a. It includes the trachea, pharynx, larynx and true vocal cords
   b. It is void of mucous.
   c. It helps keep microorganisms and dust from reaching the air sacs in the lungs
   d. It is lined externally with mucoid cilia.

_____38. Which of the following is responsible for paralysis of the cilia of the trachea?
   a. mucociliary transport
   b. tobacco smoke
   c. PSCC epithelium
   d. bacterial infection

BRONCHIAL TREE. The primary thrust of your study of this material will be in the labeling of the illustrations. This does not, however, exempt you from careful reading. Pick through the text for interesting bits and pieces. Then match the following terms with their descriptions.

_____ 39. the two tubes which split from the distal end of the trachea

_____ 40. the divisions of the bronchi within the lungs

_____ 41. divisions of the bronchi

_____ 42. divisions of the bronchioles

_____ 43. microscopic pouches in the lungs, literally the presence of a tiny cavity

_____ 44. lipid molecules secreted by cells in the epithelium of alveoli

_____ 45. condition in premature infants in which surfactant is not produced

_____ 46. the close arrangement of the epithelial wall of the alveolus, basement membrane of connective tissue, and endothelial wall of capillary

_____ 47. widespread bronchial narrowing (wheezing and coughing), perhaps caused by allergic reaction

**alveolar ducts
alveoli
asthma
bronchioles
bronchial tree
extrinsic asthma
intrinsic asthma
primary bronchi
respiratory membrane
respiratory distress
     syndrome (RDS)
surfactant**

LUNGS. Complete each of the statements by selecting the best choice. The first time you study, write the answers in the blanks. Check your answers. The second time through the questions, underline your answer without looking at your previous choices. The third time through, read only the terms in the blanks, and supply the definition from memory.

_____ 48. The (right/left) lung is somewhat thicker and broader than the other because of the position of the liver.

_____ 49. The narrow superior portion of each lung is the (apex/base).

_____ 50. The surface of the lungs that lies against the ribs is the (medial/costal) surface.

_____ 51. The surface of the lungs that faces the midline toward the heart is called the (medial/costal) surface.

_____ 52. The outer layer of serous membrane surrounding each lung is the (parietal pleura/visceral pleura).

_____ 53. The inner layer of serous membrane surrounding each lung is the (parietal pleura/visceral pleura).

_____ 54. The space between the two pleural layers is called the (pleurae/pleural cavity).

_____ 55. The (right/left) lung contains three lobes.

_____ 56. Lobes of the lungs are divided into smaller compartments known as (lobules/segments).

_____ 57. The portion of the lung which receives a single bronchiole, an arteriole, a venule, and a lymphatic vessel is a (segment/lobule).

MECHANICS OF BREATHING. Merely reading the title of this section may stimulate the reader to take a deep breath. The physiology of breathing is excellent material for an essay. To prepare, read the section thoroughly. Study the illustrations.

58. What are the two main events of pulmonary ventilation (breathing)? Write a simple definition of each.

59. In which direction do air molecules move when the pressure in a certain area is decreased?

60. During inspiration, what causes the thoracic cavity to expand?

STUDY HINT: The **external** intercostals are involved in **inspiration**, while the **internal** intercostal muscles are involved in **forced expiration**.

61. Place the following events in the correct sequence to describe inspiration. Write out your answer. This is excellent preparation for an essay type question.

   a. Alveolar pressure decreases below atmospheric levels.
   b. Air rushes into the alveoli.
   c. The diaphragm and external intercostal muscles contract.
   d. The lung surface is pulled outward, causing the lung volume to increase.
   e. Pleural cavity pressure decreases.
   f. The thoracic cavity expands.

62. Contrast inspiration and expiration using the analogies of a vacuum cleaner and a rubber band recoiling.

63. What prevents the total collapse of alveoli after expiration?

64. Place the following events in the correct sequence to describe expiration. Write out your answer as an essay.

    a. The diaphragm and external intercostal muscles relax.
    b. Air flows out of the alveoli.
    c. Alveolar pressure becomes greater than atmospheric pressure.
    d. Pleural cavity pressure increases.
    e. The thoracic cavity decreases in size.

65. What is "forced expiration"?

RESPIRATORY VOLUMES. Match these terms with their descriptions. Answers may be used more than once or not at all.

_____ 66. instrument used to measure respiratory volume

_____ 67. volume of air which moves into and out of the lungs during normal, healthy, quiet breathing in an adult

_____ 68. the amount of air that can be inhaled forcibly over the tidal volume

_____ 69. maximum amount of air that can be forcibly exhaled after a tidal expiration

_____ 70. volume of air remaining in lungs after a forced expiration

_____ 71. total amount of exchangeable air

_____ 72. sum of the vital capacity and the residual volume

_____ 73. air entering the respiratory tract without reaching the alveoli

anatomic dead space volume
ERV
expiratory reserve volume
inspiratory reserve volume
IRV
residual volume
RV
spirometer
tidal volume
TLC
total lung capacity
TV
TV + IRV + ERV = VC
vital capacity

MEMORIZE THE AVERAGE RESPIRATORY VOLUMES ON PAGE 851. Write them, from memory, here:

EXCHANGE OF GASES. Read this section aloud. If your attention needs further focusing, use a pointer or card to pace your reading. Answer the following multiple choice questions by selecting the best of the choices given.

_____74. The exchange of gases between the alveoli of the lungs and the blood involves
   a. osmosis of oxygen
   b. diffusion of oxygen and osmosis of carbon dioxide
   c. diffusion of oxygen and carbon dioxide
   d. forced diffusion against the concentration gradient

_____75. The air we breathe is a mixture of several gases including
   a. about 21% oxygen
   b. about 90% nitrogen
   c. about 80% oxygen
   d. about 20% carbon dioxide

_____76. Which of the following statements is false?
   a. The amount of pressure N, O, and CO_2 each create is called its partial pressure.
   b. If two gases dissolve, each gas diffuses in response to its own partial pressure.
   c. The rate of diffusion of oxygen is determined by the percentage of nitrogen.
   d. Each gas diffuses independently of the other gases until its partial pressure becomes equalized in the two regions.

_____77. External respiration is
   a. the exchange of gases between the alveoli and bloodstream.
   b. the inhalation of gases from outside the body into the nose or mouth.
   c. the exhalation of gases from the body to outside the body.
   d. both the inhalation and exhalation of gases through the nose or mouth.

_____78. Where is the partial pressure of oxygen greatest?
   a. in the blood plasma
   b. in the alveoli

_____79. Which direction will oxygen diffuse?
   a. from the blood plasma to the alveoli
   b. from the alveoli to the blood plasma
   c. both directions simultaneously

_____80. Oxygen-filled hemoglobin is referred to as
   a. red blood
   b. blue blood
   c. oxyhemoglobin
   d. iron rich blood

_____81. Where is the partial pressure of carbon dioxide greatest?
   a. in the blood plasma
   b. in the alveoli

_____82. Which direction does carbon dioxide diffuse?
   a. from the blood plasma to the alveoli
   b. from the alveoli to the blood plasma
   c. both directions simultaneously

_____83. Which of the following is true of internal respiration.
   a. Internal respiration is the exchange of gases between body cells.
   b. It involves the movement of oxygen from the interstitial fluid to capillaries.
   c. It involves the movement of carbon dioxide from cells into the interstitial fluid, and into the capillaries.
   d. It differs from external respiration in that it does not rely on a pressure gradient.

_____84. Carbon dioxide is mostly carried in the blood as
   a. dissolved CO_2 gas
   b. bicarbonate ion in the plasma
   c. bicarbonate ion in the red blood cells
   d. carbaminohemoglobin

_____85. CO_2 is converted to HCO_3^- in the
   a. plasma
   b. red blood cells
   c. alveoli

_____86. The enzyme which converts CO_2 into bicarbonate is
   a. hemoglobin
   b. carbonic anhydrase

CONTROL OF BREATHING. As you read, match these terms with their descriptions.

_____ 87. group of neurons in the brain stem (medulla and pons) that controls breathing

_____ 88. part of the respiratory center containing dorsal and ventral respiratory groups that control the basic rhythm of tidal breathing

_____ 89. part of the respiratory center that changes basic rhythm

_____ 90. part of the respiratory center that transmits impulses that inhibit expiration

_____ 91. literally, to move toward the lung

_____ 92. literally, not breathing

_____ 93. located in the medulla, this monitors CO_2 levels in the blood of the brain

_____ 94. located in the aorta and carotid arteries, these monitor levels of O_2, CO_2, and H^+

_____ 95. increased rate of ventilation by the respiratory center to "blow off" excess carbon dioxide

apneustic
apneustic area in the pons
bends
chemosensitive area
chemoreceptors
hyperventilation
medullary rhythmicity center
pneumotaxic
pneumotaxic area in the pons
rapture of the deep
respiratory center
SCUBA
stretch receptors

_____ 96. located in walls of bronchi and bronchioles, these regulates the degree of stretch in the lungs

_____ 97. self contained underwater breathing apparatus

_____ 98. increase in nitrogen levels in the brain during deep sea diving

_____ 99. accumulation of bubbles of nitrogen due to rapid ascent from diving

apneustic
apneustic area in the pons
bends
chemosensitive area
chemoreceptors
hyperventilation
medullary rhythmicity center
pneumotaxic
pneumotaxic area in the pons
rapture of the deep
respiratory center
SCUBA
stretch receptors

HOMEOSTASIS. After reading the two paragraphs describing the efficiency of the respiratory system in maintaining homeostasis, answer the following question.

100. How does the respiratory system maintain homeostasis during a strenuous exercise session when the demands for oxygen increase dramatically? Be sure to correctly use and define "hyperpnea" in your answer.

CLINICAL TERMS OF THE RESPIRATORY SYSTEM. Using the correctly spelled clinical term, fill in the blanks in these statement which describe respiratory problems.

_____ 101. a chronic disease of the alveoli in which their walls are destroyed and breathing is extremely laborious

_____ 102. difficulty in breathing

_____ 103. a respiratory disease characterized by fluid accumulation in the lungs, wheezing, and shortness of breath

_____ 104. a temporary absence of breathing

_____ 105. deficiency of oxygen within tissues

_____ 106. a condition in which there is an excess of carbon dioxide in the blood and a deficiency of oxygen

_____ 107. inherited disorder with thick, sticky mucus, lung damage, disturbed digestion, and unusual sweating and salivating

_____ 108. blood in the pleural cavity

_____ 109. air in the pleural cavity causing collapse of the lung

_____ 110. when a blood clot moves in the lungs and causes blockage of blood flow

_____ 111. viral, bacterial, of fungal infection of the alveoli resulting in fluid in the alveoli

_____ 112. chronic lung disease resulting in the reduction of respiratory membrane surface, caused by exposure to coal or cigarette smoke

_____ 113. bronchogenic carcinoma affecting smokers more than non-smokers

_____ 114. reduced availability of oxygen in the blood

_____ 115. bacterial infection of the tonsils

_____ 116. surgical incision through trachea to clear air passageway

_____ 117. crib death due to cessation of breathing in infants

_____ 118. inflammation of the nasal cavity lining

_____ 119. inflammation of the sinus cavity lining

_____ 120. for example, the common cold, flu, pharyngitis, laryngitis, and bronchitis

LABEL AND LIST

1. List three factors that influence the rate and depth of breathing (other than the control exerted by the respiratory center).

    a.   b.

    c.

2. Label the respiratory tract.

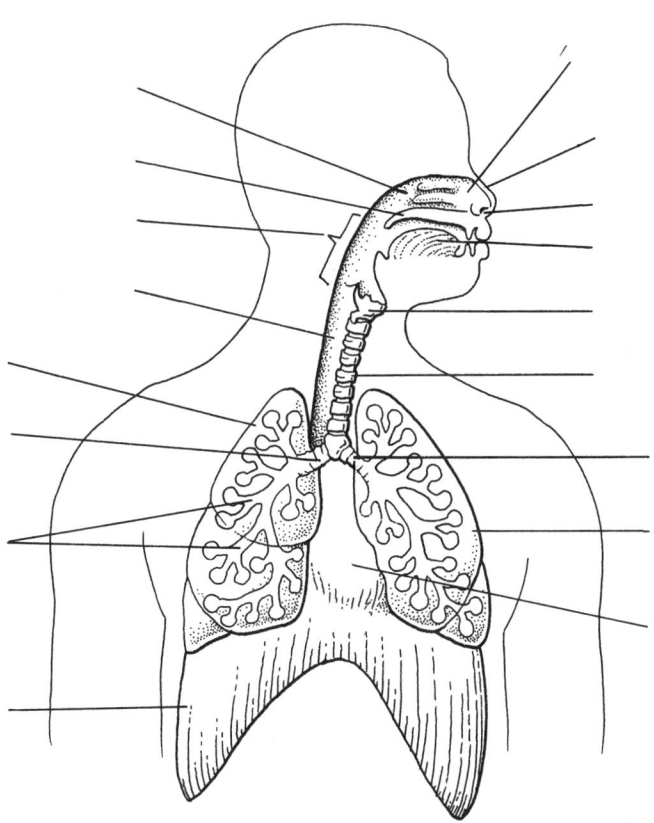

303

3. Label the structures of the upper respiratory tract.

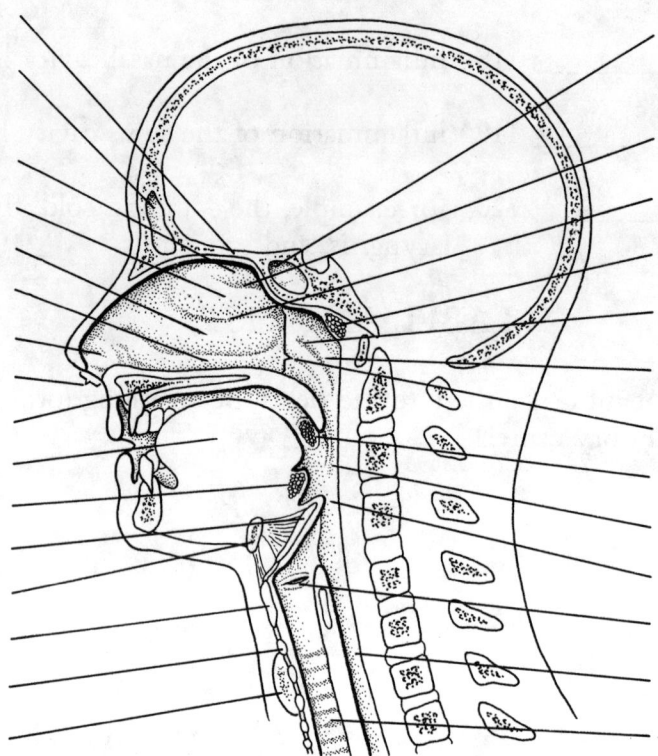

4. Label the larynx.

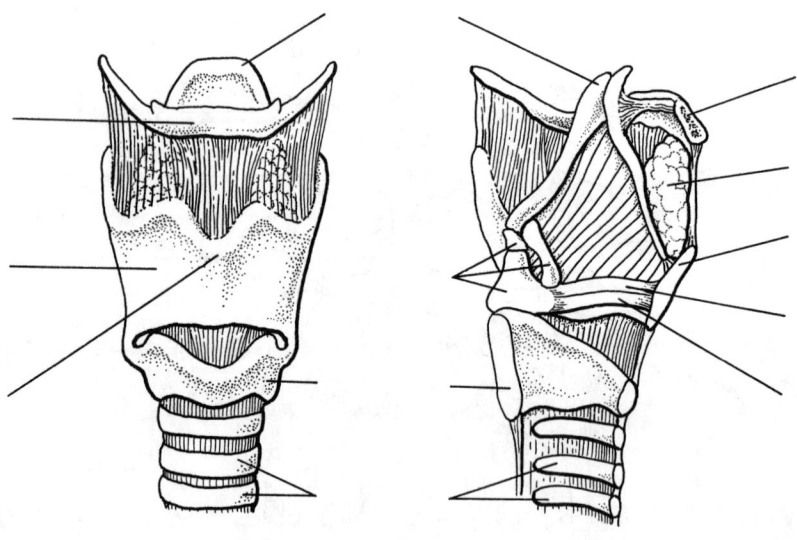

ADDITIONAL STUDY

Read the Chapter Summary (pages 466 - 467). Write out the definitions of all the KEY TERMS (page 467).

Review the illustrations in your textbook and understand the answers to the questions associated with each one. The answers are on pages 468 - 469.

Having studied this chapter, close your book, put away your notes, and test yourself by **writing** the answers to the "CONCEPTS CHECKS" and "QUESTIONS FOR REVIEW" in your text. Writing the answers will force you to challenge yourself. If you can write the answers for yourself, you will have more confidence in writing answers for your teacher.

The day before an exam over this chapter, read the "Learning Objectives", page 442, and review any of the sections which you think might cause a problem.

THE RESPIRATORY SYSTEM
Robert W. Bauman, Jr., Ph.D.

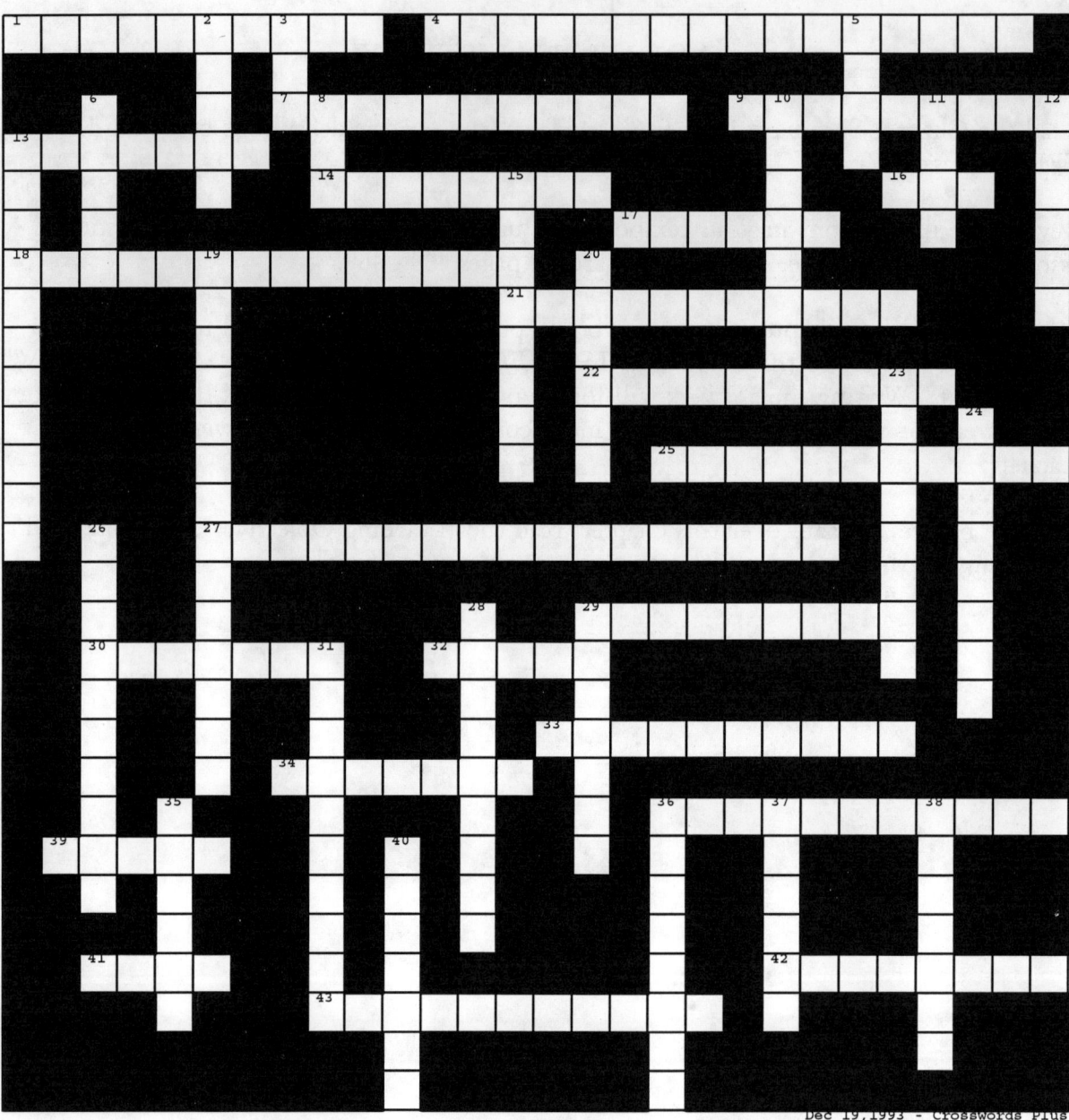

Dec 19, 1993 - Crosswords Plus

Across

1. breathing out
4. increase in the rate of ventilation
7. breathing in and out
9. neurons which inhibit expiration
13. windpipe
14. portion of pleura adjoing the lungs
16. results from lack of surfactant in infants
17. deficiency of oxygen in tissues
18. inherited condition of thickened mucus
21. breathing in
22. oxygen carrying chemical
25. region of pons which inhibits the breathing rate
27. enzyme which produces carbonic acid
29. infection of the alveoli
30. opening of the trachea
32. total capacity of exchangeable air
33. counteracts surface tension of water within the alveoli
34. branches of the trachea
36. gas exhange between the external environment and cells
39. volume of air moving in and out of the lungs during normal quiet breathing
41. largest respiratory organ
42. small sacs where external respiration occurs
43. center in the medulla which sets normal breathing rate

Down

2. short cessation of pulmonary ventilation
3. maximum amount of air that can be inhaled above the tidal volume
5. superior portion of lung
6. external openings of nasal cavity
8. maximum amount of air that can be forcibly exhaled over tidal volume
10. portion of pleura that lines the thoracic cavity
11. crib death
12. associated with smoking
13. incision through trachea
15. amount of air which cannot be forcibly exhaled
19. receptors which respond to oxygen, carbon dioxide, and acidity
20. respiratory disease which can allergic or idiopathic
23. type of respiration which involves exchange of gases between the blood and body cells
24. type of respiration involving alveoli and capillaries
26. cartilage which blocks the glottis
28. primary muscle of ventilation
29. cavity between pleurae
31. device to measure ventilation
35. voicebox
36. inflammation of the nasal cavity
37. two layers of serous membrane surrounding the lungs
38. cartilage making the 'Adam's apple'
40. tube connecting the nasal cavity, oral cavity, and larynx

CHAPTER 16: THE DIGESTIVE SYSTEM

You have heard the phrase, "You are what you eat." In a real sense, food becomes a part of us. How that food is processed to a size that is useable by the cells for metabolism, growth, and reproduction, is the task of the digestive system. You should "devour" this chapter as it is full of "delicious" information to "sink your teeth into." "Bon Appetit!"

CONTENT MASTERY

ORGANIZATION. After reading the short description of the organization of the digestive system, answer the following.

1. The two main categories of digestive organs are the ... and the ...

2. The alimentary canal is also known as the ... tract.

3. The alimentary canal averages ... feet in length.

4. The ... organs, such as the teeth, pancreas, and liver, assist the organs of the GI tract.

DIGESTIVE PROCESSES. Note the six most important digestive processes. Match them and a few additional vocabulary words with their descriptions. Pay attention to spelling.

_____ 5. process of bringing food into the digestive system

_____ 6. movement of food through the GI tract

_____ 7. movement of food through the GI tract by a series of contractions

_____ 8. series of muscular events that push food through the mouth and throat

_____ 9. breakdown of food particles by the human machine (chewing, mixing, churning)

absorption
chemical digestion
defecation
ingestion
mastication
mechanical digestion
peristalsis
propulsion
swallowing

_____ 10. chewing

_____ 11. breakdown of large
molecules into their basic
chemical building blocks

_____ 12. transport of digested food
material from the lumen
of the GI tract to the
blood or lymph

_____ 13. elimination of indigestible
material from the body
as feces

absorption
chemical digestion
defecation
ingestion
mastication
mechanical digestion
peristalsis
propulsion
swallowing

SPECIAL FEATURES OF THE DIGESTIVE SYSTEM. Consider the peritoneum and the similarity of structure of most organs of the alimentary canal. Define the following as you read, and locate them on Figure 16-4.

14. Peritoneum -

15. Parietal peritoneum -

16. Serosa -

17. Peritoneal cavity -

18. Peritoneal folds -

19. Falciform ligament -

20. Lesser omentum -

21. Mesentery -

22. Greater omentum -

23. Mucosa -

24. Submucosa -

25. Muscularis -

26. Serosa -

DIGESTIVE ORGANS. Much of your study of the anatomy of these organs will be associated with labeling the diagrams in the LABEL AND LIST section of the study guide. Additional "tidbits" will be studied here. Enjoy these "tasty morsels" of information.

MOUTH.   Match these terms associated with the mouth with their descriptions.

_____ 27. chewing

_____ 28. responsible for beginning chemical digestion

_____ 29. bony, anterior roof of the oral cavity

_____ 30. opening of the oral cavity into the pharynx

_____ 31. finger-like projection which prevents food from entering the nasal cavity

_____ 32. literally, entry chamber

_____ 33. anchors the tongue to the floor of the mouth

_____ 34. provide friction for handling food

_____ 35. literally, "little bridle", pertaining to the tongue

cheeks
fauces
hard palate
lingual frenulum
lips
mastication
oral cavity
papillae
saliva
soft palate
uvula
vestibule

TEETH. Match these terms with their descriptions.

_____ 36. number of deciduous teeth

_____ 37. number of permanent teeth in a "full set"

_____ 38. teeth that are chisel shaped

_____ 39. cone-shaped teeth useful for tearing

canines
crown
dental caries
dentin
enamel
gingivitis
incisors
molars
periodontal ligament
thirty two
twenty four
twenty

_____ 40. flattened teeth useful in crushing

_____ 41. the visible part of the tooth

_____ 42. hardest substance in the body

_____ 43. anchors a tooth within the jaw

_____ 44. tooth decay

_____ 45. tender, bleeding gums

_____ 46. part of tooth covered with enamel

canines
crown
dental caries
dentin
enamel
gingivitis
incisors
molars
periodontal ligament
thirty two
twenty four
twenty

SALIVARY GLANDS AND DIGESTION IN THE MOUTH. Select the choice which best completes the following statements.

_____ 47. The composition of saliva is
   a. 99.5% water and 0.5% solutes
   b. 99.5% solutes and 0.5% water
   c. 75.5% water and 24.5% enzymes and other solutes
   d. none of the above

_____ 48. What is the purpose of saliva?
   a. It dissolves food molecules.
   b. It is necessary for taste.
   c. The mucus in saliva eases swallowing.
   d. All of the above are correct.

_____ 49. The largest salivary glands are
   a. parotid
   b. submandibular
   c. sublingual
   d. buccal

_____ 50. The salivary gland whose name means "beside the ear" is the
   a. parotid gland.
   b. submandibular gland.
   c. sublingual gland.
   d. buccal gland.

_____51. The salivary gland whose name means "below the lower jaw" is the
   a. parotid gland.
   b. submandibular gland.
   c. sublingual gland.
   d. buccal gland.

_____52. The salivary gland whose name means "below the tongue" is the
   a. parotid gland.
   b. submandibular gland.
   c. sublingual gland.
   d. buccal gland.

_____53. Which of the following represents mechanical digestion?
   a. mastication and mixing
   b. salivary amylase and a bolus
   c. chewing and the production of saliva
   d. the splitting of starch and glycogen molecules

_____54. Which of the following is NOT true of a bolus?
   a. A bolus is formed in the mouth.
   b. A bolus is coated with mucus.
   c. A bolus is the food in the mouth which is completely digested.
   d. A bolus is formed as a result of mastication in the presence of saliva.

PHARYNX AND ESOPHAGUS. Fill in the blanks with the appropriate terms as you read about the pharynx and esophagus.

bolus
epiglottis
esophageal hiatus
esophagus
heartburn
internal nares
laryngopharynx
larynx
lower esophageal sphincter
nasopharynx
oropharynx
peristalsis
pharynx
soft palate
ten
tongue
twenty

55. pharynx
56. internal nares
57. larynx
58. nasopharynx
59. oropharynx
60. laryngopharynx
61. bolus
62. tongue
63. soft palate
64. pharynx
65. epiglottis
66. peristalsis
67. esophagus
68. ten
69. esophageal hiatus
70. lower esophageal sphincter
71. heartburn

The (55) is the chamber located behind the oral cavity extending from the (56) to the (57). It is divided into three segments, the (58), the (59), and the (60).

Swallowing begins as the food (61) is pushed from the oral cavity into the pharynx by the (62). As this occurs, the (63) rises to prevent food from entering the nasal cavity. This is followed by the contraction of muscles in the wall of the (64) which move the larynx upward as the (65) presses downward closing off the airway and opening the esophagus. The bolus moves onward to the stomach by (66). This process continues along the length of the esophagus.

The (67) is a muscular tube that extend from the pharynx to the stomach for about (68) inches. It penetrates the diaphragm through an opening called the (69). Near its union with the stomach, the esophagus wall is thickened to form a sphincter called the (70). When this muscle is weak, gastric juices leak back into the esophagus producing a painful condition called (71).

STOMACH. Fill in the blanks with the appropriate terms concerning the anatomy of the stomach.

_____ 72. The deep folds in an empty stomach are known as ...

_____ 73. The convex lateral margin of the stomach is called the ...

_____ 74. The concave medial margin is known as the ...

_____ 75. At the terminal end of the pylorus is a muscle called the ..

_____ 76. "Fundus" literally means ...

_____ 77. "Pylorus" literally means...

_____ 78. The millions of microscopic openings of the mucosa of the stomach are called ...

_____ 79. Gastric juice is secreted by ...

_____ 80. The gastric glands contain a variety of secretory cells, including ... cells that secrete digestive enzymes

_____ 81. The cells that secrete hydrochloric acid in the stomach are ... cells.

_____ 82. The ... cells secrete mucus.

_____ 83. There is both ... and chemical digestion in the stomach.

Using Table 16-1 and the sections that describe the "Functions of the Stomach", match these terms with their descriptions. Terms may be used more than once or not at all.

_____ 84. produced by zymogenic cells of the stomach

_____ 85. produced by parietal cells of the stomach
_____ (two answers)

_____ 86. produced by the mucous cells of the stomach

_____ 87. an inactive form of pepsin

_____ 88. provides an acidic environment that is needed for pepsin activation

_____ 89. provides a protective layer of mucus along the mucosal lining

_____ 90. enables the absorption of B12 across the small intestinal wall

_____ 91. the most important enzyme in the gastric juice

_____ 92. hole in the stomach wall

_____ 93. a semi-fluid paste of food and gastric juice

_____ 94. regulate the activities of the stomach (two
_____ answers)

chyme
cholecystokinin
gastric ulcer
gastrin
HCl
hormones
involuntary control centers of the brain
intrinsic factor
mucus
mucous cells
parietal cells
pepsin
pepsinogen
secretin
zymogenic cells

_____ 95. hormone which stimulates secretion of gastric juice and peristalsis of the stomach

_____ 96. intestinal hormones which inhibit stomach peristalsis but stimulate other actions

chyme
cholecystokinin
gastric ulcer
gastrin
HCl
hormones
involuntary control
   centers of the brain
intrinsic factor
mucus
mucous cells
parietal cells
pepsin
pepsinogen
secretin
zymogenic cells

PANCREAS. Read the text material which describes the pancreas. Whenever there is a diagram, pay close attention. Memorize structures as you study each diagram. Then match these terms with their descriptions. Terms may be used more than once. When there are more than one answer parenthesis are used to denote the number of answers.

_____ 97. exocrine secretory cells in the pancreas

_____ 98. tube originating from the liver which usually fuses with the pancreatic duct

_____ 99. first segment of the small intestine

_____ 100. cells scattered among the acini that produce hormones

_____ 101. pancreatic enzyme that digests carbohydrates

_____ 102. splits starch and glycogen molecules into maltose subunits

acini
carboxypeptidase
cholecystokinin
chymotrypsin
common bile duct
duodenum
endocrine
nucleases
pancreatic juice
pancreatic lipase
pancreatic amylase
secretin
trypsin

_____ 103. pancreatic enzyme that
_____   digests proteins
_____   (three answers)

_____ 104. pancreatic enzyme that aids
                             in the digestion of fats

_____ 105. pancreatic enzyme that
                             splits fat into fatty acids
                             and glycerol

_____ 106. pancreatic enzymes that
                             digest nucleic acids into
                             nucleotides

_____ 107. hormones secreted by the
                             small intestine that
_____   regulate pancreatic
                             secretions (two answers)

_____ 108. when HCl is detected in
                             the small intestine, this
                             hormone is released so
                             that chyme is neutralized

acini
carboxypeptidase
cholecystokinin
chymotrypsin
common bile duct
duodenum
endocrine
nucleases
pancreatic juice
pancreatic lipase
pancreatic amylase
secretin
trypsin

LIVER. Take a deep breath (increasing the oxygen supply to your brain), sit up straight, and read the section on the liver. Then select the term in parenthesis which best completes each of the following. Write your answer in the blanks.

_____ 109. The liver is divided into (one/two/three) main
                             sections called lobes.

_____ 110. The right lobe is (larger/smaller) than the left lobe.

_____ 111. The structural and functional subunits of the liver
                             are (liver lobules/liver lobes).

_____ 112. Liver cells are known as (hepatocytes/Kupffer
                             cells/sinusoids).

_____ 113. Blood from both the hepatic portal vein and the hepatic artery passes through channels called (hepatocytes/central veins/sinusoids).

_____ 114. Phagocytic cells which remove bacteria from the blood are (hepatocytes/Kupffer cells/canaliculi).

_____ 115. Bile is secreted by (hepatocytes/sinusoids/Kupffer cells).

_____ 116. The (bile salts/water/bilirubin/cholesterol) of bile function(s) in fat digestion

_____ 117. The (hepatic/cystic/pancreatic) duct unites with the cystic duct from the gall bladder to form the common bile duct.

_____ 118. The digestive role of the liver is the secretion of (electrolytes/cholesterol/bile).

_____ 119. The process of breaking apart clumps of fat molecules into tiny droplets is called (lipid mastication/emulsification).

_____ 120. Under the direction of hormones released by the pancreas, the liver may store glucose as (disaccharide/maltose/glycogen).

_____ 121. The liver's fat-transporting proteins that transport triglycerides to fat tissue for storage are (LDLs/VLDLs/HDLs).

_____ 122. (LDLs/HDLs/VLDLs) transport cholesterol to cells outside the liver for membrane or hormone synthesis.

_____ 123. Cholesterol is transported by (HDLs/LDLs/VLDLs) to the liver where it becomes part of bile.

_____ 124. The lipoprotein that can cause atherosclerosis by accumulating in blood vessels are (LDLs/HDLs/VLDLs).

_____ 125. The scarring of the liver from any cause is referred to as (multiple sclerosis/cirrhosis of the liver).

_____ 126. Scarring of the liver is (reversible/irreparable).

_____ 127. The most important role of the liver is in (bile production/protein metabolism)

_____ 128. Excess amino acids are converted into (cholesterol/bile/urea) by the liver.

_____ 129. The various functions of the liver, with the exception of phagocytosis, are performed by (Kupffer cells/hepatocytes).

_____ 130 Toxins are (detoxified/ignored) by the liver.

STUDY TIP: Study the table 16-2 which summarizes the many functions of the liver. Learn it well enough that you can accurately describe the specific roles of the liver when the right column is covered. Be able to list the six major functions of the liver given in the left column. (You will be asked to list them again in the LABEL AND LIST section .)

GALLBLADDER. As you read this short section about the small, thin-walled sack called the gallbladder, answer the following questions.

131. What is the primary function of the gallbladder?

132. Describe the ducts that make up the common bile duct.

133. What is the small sphincter muscle at the end of the common bile duct? When does it open to allow bile to enter the small intestine?

SMALL INTESTINE. As you read about the small intestine, fill in the following blanks using a pencil. This exercise will thus be a reading pacer and will serve to help focus your attention. When you have completed your work and corrected your answers, erase them and test yourself.

_____ 134. The body's most important digestive organ is the ...

_____ 135. The alimentary canal's longest segment is the ... which is about ... feet long.

_____ 136. The three segments of the small intestine are the ..., the ..., and the ...

_____ 137. ... literally means "twelve finger-breadths in length."

_____ 138. ... literally means "empty."

_____ 139. ... literally means "flank or groin."

_____ 140. The connection between the large intestine and the small intestine is the ... valve

_____ 141. The small intestine is noted for its ability to absorb food and to complete ... digestion.

_____ 142. Projections of the small intestine mucosa are ...

_____ 143. The microscopic projections on each villus are ...

_____ 144. The effect of microvilli and villi is to increase the .... of the intestinal lining.

_____ 145. After nutrients are absorbed into a villus, they are carried away via capillaries and a lymphatic vessel called a ...

_____ 146. "Villi" literally means ...

_____ 147. ... glands are responsible for secreting water and mucus into the intestine.

_____ 148. ... of the submucosa protect the body against infections microorganisms.

_____ 149. The glands which secrete alkaline mucus to neutralize the acidic chyme are called ...

_____ 150. The folds of the mucosa and submucosa are called ...

_____ 151. The microvilli of the intestinal mucosa contain packets of intestinal enzymes. (true/false)

_____ 152. Condition resulting from the enzyme lactamase is ...

_____ 153. It takes between ... and ... hours to move chyme all the way through the small intestine.

NOTE: The last portion of the small intestine is the ILEUM. The superior portion of the hip bone is the ILIUM (chapter 7). Note the spellings!

LARGE INTESTINE. Fill in the blanks using the terms listed. Terms may be used more than once or not at all. After you have completed this exercise and corrected your work, test yourself by filling in the blanks without the aid of the list of terms.

_____ 154. the final segment of the alimentary canal

_____ 155. length of large intestine in feet

_____ 156. diameter of large intestine in inches

_____ 157. literally, "blind pouch"

_____ 158. literally, "worm-like attachment"

_____ 159. literally, "pertains to the large intestine"

_____ 160. literally, "ring"

_____ 161. infection of appendix

20
3
5
anus
anal canal
anal columns
appendicitis
bile pigments
cecum
colon
defecation
feces
fiber
haustrum
intestinal flora
peritonitis
rectum
taenia coli
vermiform appendix
vitamin K

_____ 162. infectious material in peritoneal cavity resulting in life-threatening condition

_____ 163. external opening of the large intestine

_____ 164. parallel folds of the anal canal that reduce friction with feces during defecation

_____ 165. bands in the muscularis of the large intestine

_____ 166. pouch of the large intestine resulting from contraction of smooth muscle bands

_____ 167. responsible for the color of feces

_____ 168. produce vitamin K in large intestine

_____ 169. responsible for aromatic gases in feces

_____ 170. the propulsion of feces out the anus

_____ 171. time (in hours) that it takes for material to pass through the colon

_____ 172. In a daily diet, this helps to reduce the chances of developing diverticulosis later in life.

20
3
5
anus
anal canal
anal columns
appendicitis
bile pigments
cecum
colon
defecation
feces
fiber
haustrum
intestinal flora
peritonitis
rectum
taenia coli
vermiform appendix
vitamin K

HOMEOSTASIS. Carefully read the section which describes the role of the digestive system in supporting homeostasis before answering the following question.

173. Why does interstitial fluid collect in the extracellular spaces when homeostasis of the digestive system is disrupted by malnutrition?

CLINICAL TERMS OF THE DIGESTIVE SYSTEM. Match the terms with their descriptions.

_____ 174. inflammation of the diverticula in the wall of the large intestine

_____ 175. uptight smooth muscles between stomach and small intestine

_____ 176. inflammation and death of liver tissue

_____ 177. enlarged veins in the lining of the anal canal

_____ 178. *Candida albicans* in the mouth of infants

_____ 179. lack of hydrochloric acid in gastric juice resulting in insufficient digestion of protein

_____ 180. inability to swallow

_____ 181. inflammation of gallbladder

**achalasia
achlorhydria
aphagia
cancers of digestive organs
cholecystitis
colitis
diverticulitis
dysentery
enteritis
gastric ulcer
gastritis
hemorrhoids
hepatitis
hiatal hernia
intestinal ulcer
pancreatitis
peptic ulcer
thrush**

_____ 182. inflammation of colon and rectum

_____ 183. protrusion of stomach through the diaphragm and into the thoracic cavity

_____ 184. inflammation of the stomach

_____ 185. Chron's disease

_____ 186. digestion of the pancreas, often caused by excessive alcohol consumption

_____ 187. excessive acid secretions and digestive enzymes which destroy mucosal lining

_____ 188. peptic ulcer in the stomach

_____ 189. peptic ulcer in the small intestine

_____ 190. carcinomas of the alimentary canal

_____ 191. severe diarrhea and cramps caused by bacteria, viruses, or protozoans

achalasia
achlorhydria
aphagia
cancers of digestive organs
cholecystitis
colitis
diverticulitis
dysentery
enteritis
gastric ulcer
gastritis
hemorrhoids
hepatitis
hiatal hernia
intestinal ulcer
pancreatitis
peptic ulcer
thrush

LABELS AND LISTS

1. List the six most important processes accomplished by the digestive system.

   a.   b.

   c.   d.

   e.   f.

2. List four folds of the parietal peritoneum.

   a.   b.

   c.   d.

3. What are the four distinct tissue layers of the wall structure of the organs of the alimentary canal?

   a.   b.

   c.   d.

4. What are the three accessory organs of the mouth?

   a.   b.

   c.

5. List four types of teeth in an ordinary, healthy mouth.

   a.   b.

   c.   d.

6. What four types of glands produce saliva?

   a.   b.

   c.   d.

7. Into what three segments is the pharynx divided?

   a.   b.

   c.

8. List four regions of the stomach.

   a.   b.

   c.   d.

9. List four layers of the stomach wall.

   a.   b.

   c.   d.

10. List five components of the gastric juice.

    a.   b.

    c.   d.

    e.

11. List five functions of the stomach.

    a.   b.

    c.   d.

    e.

12. What six materials are absorbed across the stomach lining?

    a.   b.

    c.   d.

    e.   f.

13. List six functions of the liver.

    a.   b.

    c.   d.

    e.   f.

14. List four causes of cirrhosis of the liver.

    a.   b.

    c.   d.

15. List three function of the small intestine.

    a.   b.

    c.

16. List two functions of the large intestine.

    a.   b.

17. What are the four main segments of the large intestine?

    a.   b.

    c.   d.

18. What are the four regions of the colon?

    a.   b.

    c.   d.

19. List five causes of malnutrition.

   a.   b.

   c.   d.

   e.

ESSAY QUESTION: Trace the digestion of a cheeseburger as it passes through the alimentary canal. Be sure to discuss mechanical and chemical digestion of lipids, carbohydrates, proteins, and nucleic acids, absorption, and defecation.

20. Label the major organs of the digestive system

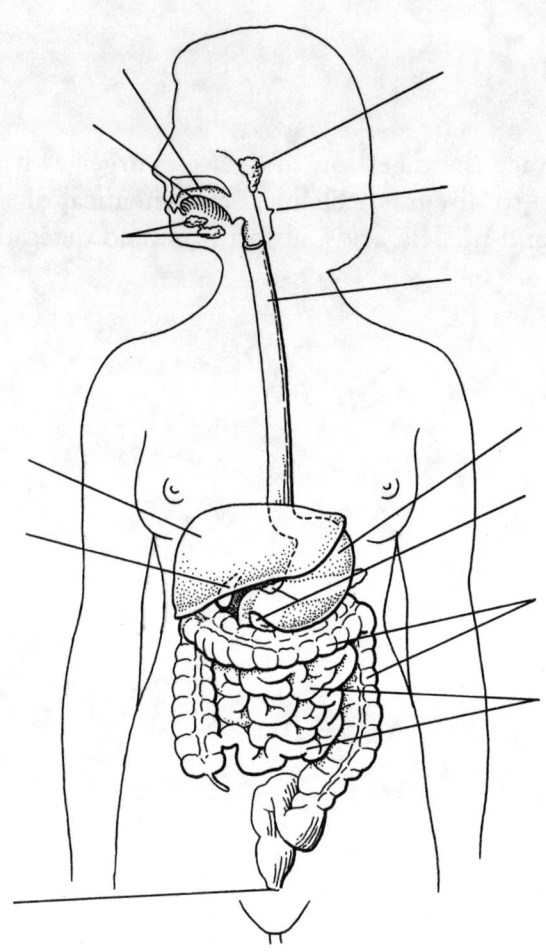

21. Label this drawing of a sagittal section of the head.

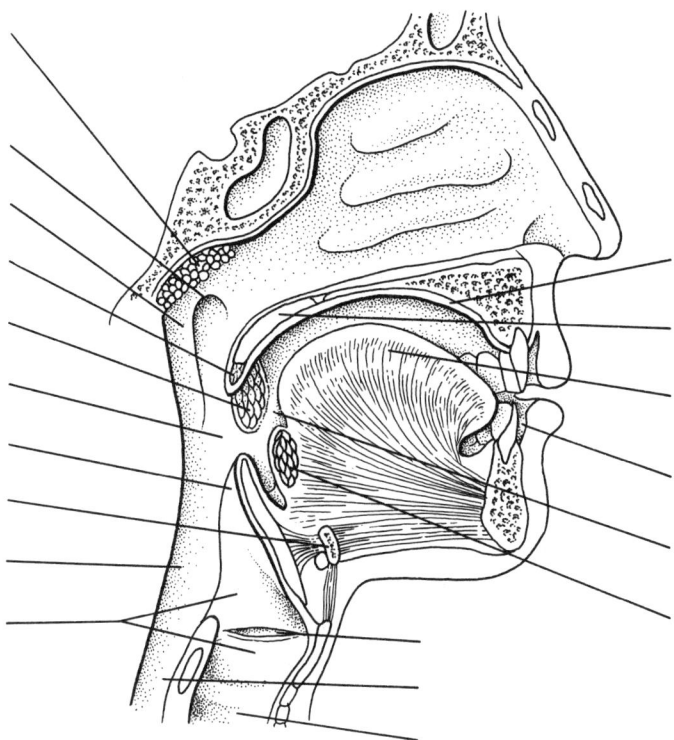

22. Label the mouth.

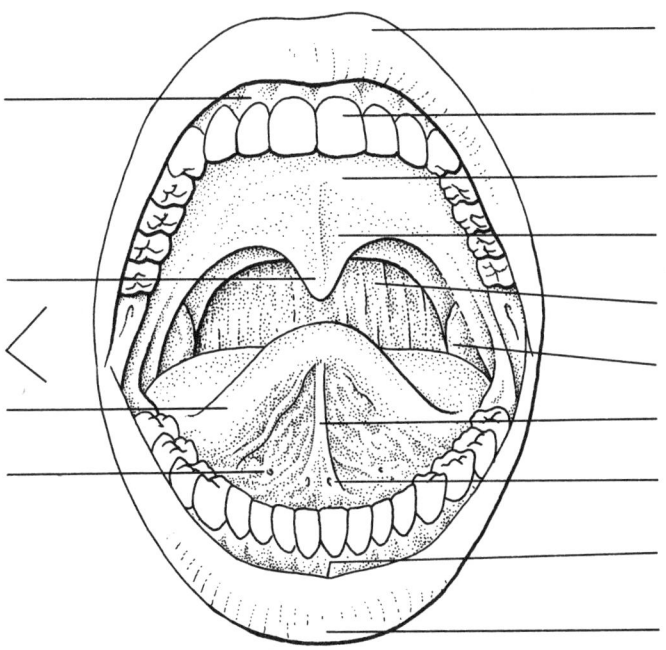

329

23. Label this tooth.

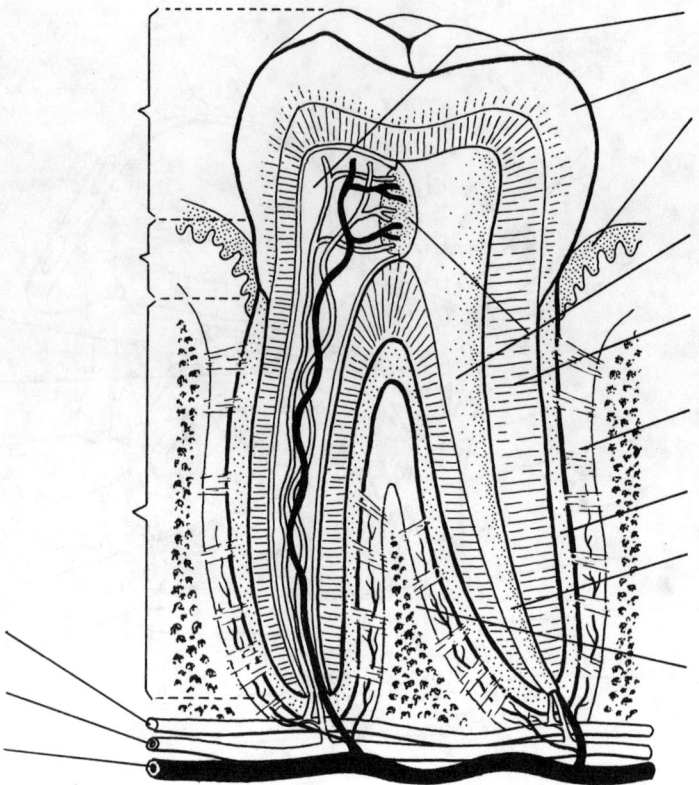

24. Label the salivary glands.

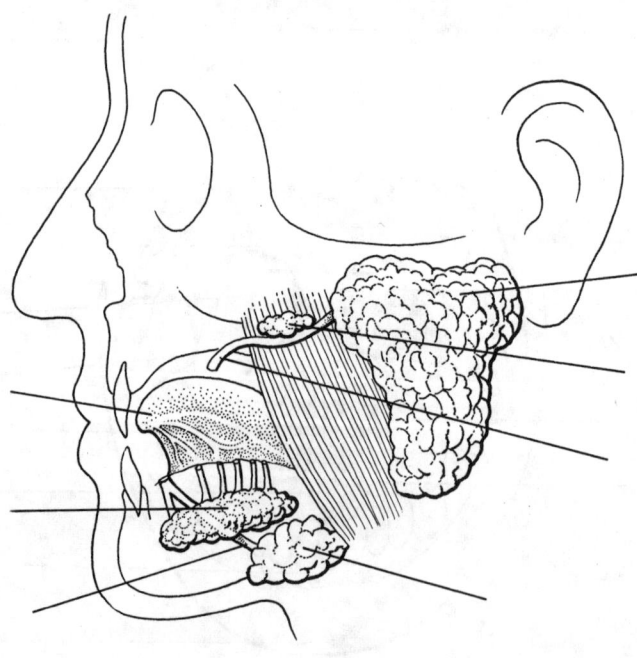

25. Label these digestive organs.

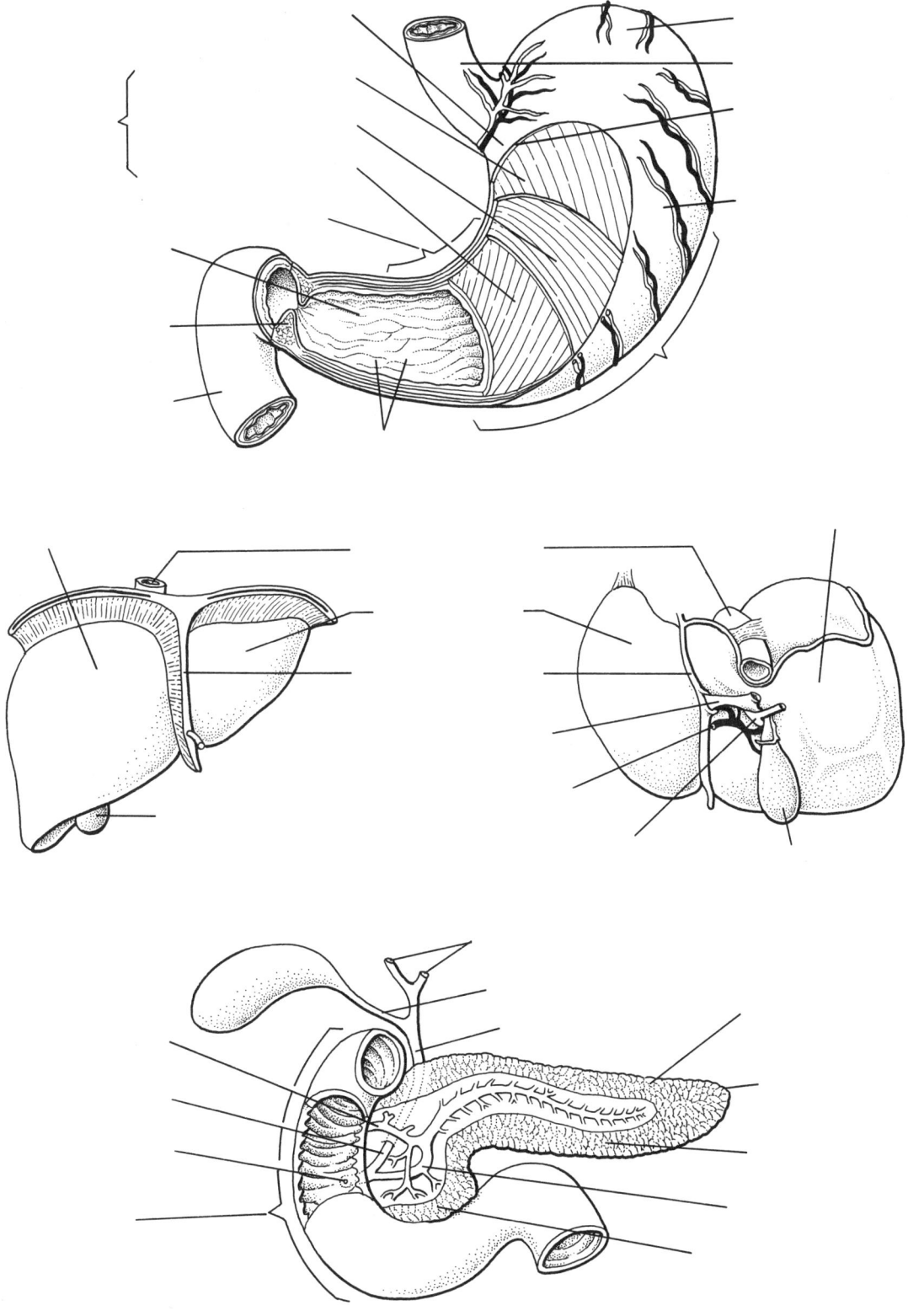

331

26. Label the large intestine.

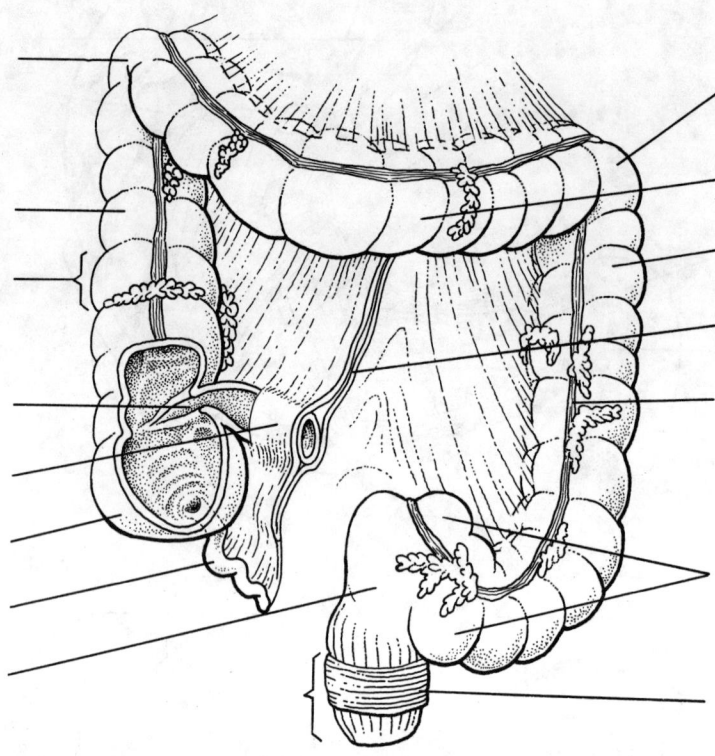

ADDITIONAL STUDY

Read the Chapter Summary (pages 502 - 504). Write out the definitions of all the KEY TERMS (page 505).

Review the illustrations in your textbook and understand the answers to the questions associated with each one. The answers are on pages 506 - 507.

Having studied this chapter, close your book, put away your notes, and test yourself by **writing** the answers to the "CONCEPTS CHECKS" and "QUESTIONS FOR REVIEW" in your text (pages 474, 477, 482, 484, 487, 489, 497, 500, and 505 - 506). Writing the answers will prepare you for an exam

The day before the exam over this chapter, read the "Learning Objectives", page 470, and review any of the sections which you think will cause you a problem.

CHAPTER 17: NUTRITION AND METABOLISM

A study of nutrition and metabolism is very timely in America as attention is being focused on healthy eating habits. Many students do not have a "mom" at home to make them eat their spinach, but we should all have an educated awareness of the role nutrition plays in good health. This is a great opportunity for you, the student, to acquire that educated awareness.

CONTENT MASTERY

NUTRIENTS. After reading the introduction to Chapter 17 and the section entitled "Nutrients," briefly answer these questions.

1. From what source does the body get energy for the multitude of chemical reactions that must take place to ensure health?

2. Once food units have been broken down by the digestive system, transported by the bloodstream to the liver and ultimately to body cells, what two functions do they have?

3. Define "nutrition."

4. Do all nutrients come from ingestion? Explain.

5. What is an essential nutrient?

CARBOHYDRATES. Indicate whether the following statements are true or false. If the statement is false, change the italicized word(s) to make it true.

_____ 6. Carbohydrates are large molecules composed of simple *lipid* subunits.

_____ 7. Starch and glycogen are known as *polysaccharides*, or complex carbohydrates.

_____ 8. Glycogen, a complex carbohydrate, can be obtained by the body *only from the liver*.

_____ 9. *Disaccharides* are readily absorbed into the bloodstream.

_____ 10. The "roughage" carbohydrate which facilitates defecation is *cellulose*.

_____ 11. People *survive* with little ingestion of carbohydrates.

FATS. "FAT FREE, "LOW FAT," "REDUCED FAT," etc. are labels which market many products today. Is all fat bad? Read about fats before matching the following. Answers may be used more than once.

_____	12. neutral fats	10%
		30%
_____	13. the most abundant source of dietary fats	40%
		cholesterol
		essential fatty acid
_____	14. contain single bonds between carbon atoms	linoleic acid
		saturated fats
		triglycerides
_____	15. contain double bonds between carbon atoms	unsaturated fats
_____	16. solid at room temperature	
_____	17. liquid at room temperature	
_____	18. type of fats found in butter	
_____	19. type of fats found in coconut oil	
_____	20. type of fats found in pecans	
_____	21. necessary fatty acid that must be ingested because the body cannot synthesize it	
_____	22. type of fat not used for energy	

_____ 23. type of fat important in the building of bile salts and steroid hormones

_____ 24. percentage of total calorie intake represented by fats as recommended by American Heart Association

10%
30%
40%
cholesterol
essential fatty acid
linoleic acid
saturated fats
triglycerides
unsaturated fats

PROTEINS. After reading about proteins, answer these questions briefly.

25. What are "essential amino acids?"

26. Approximately how many ounces of protein is required daily for a 150 pound man?

27. Contrast the apparent relationships of fat and fiber to cancer.

Now, study Table 17-1 so that you can cover any column and supply the answers.

VITAMINS. After reading about vitamins and studying Table 17-2, match the following terms with their descriptions. Terms may be used more than once or not at all. When multiple answers are required, it is indicated in parenthesis.

_____
_____ 28. molecules that assist in regulating physiological processes (2)

_____ 29. vitamins that bind to fats

_____ 30. vitamins K, A, D, and E, for example

_____ 31. the "sunshine vitamin" converted by sunlight from cholesterol in the skin

_____ 32. vitamins such as B and C

coenzymes
fat-soluble vitamins
vitamin K
vitamin A and E
vitamins
vitamin D
water-soluble vitamins

_____	33. a vitamin synthesized by bacteria in the intestine	**coenzymes**
		fat-soluble vitamins
		vitamin K
_____	34. the type of vitamin that is not destroyed by high temperatures	**vitamin A and E**
		vitamins
		vitamin D
		water-soluble vitamins
_____	35. vitamins that must be ingested regularly because they are not stored in large amounts	
_____	36. the type of vitamin that is damaged by high temperatures	

Study Tip: Table 17-2 provides a very thorough summary of vitamin requirements, the sources of these vitamins, and problems associated with too much or too little of each. Ask your professor how much of this information is important to his objectives for you in this course. As with former charts, memorize the appropriate information. Then, cover the columns, quiz yourself, and quiz yourself again until you are an expert!

Study Tip: The fat-soluble vitamins can be remembered as "Katy" - K, A, D, and E!

MINERALS. Read about minerals. Scan the chart which summarizes the sources, utilization, and problems associated with the various minerals. As you may have done with the vitamins above, ask your professor how much of this material is to be committed to memory. Answer these questions briefly.

37. Where in the body are dietary minerals mostly found?

38. Which two minerals are most important to the body in terms of structure?

39. Which minerals, for example, are prevalent in bone and teeth?

40. Which minerals are found in ionized form in body fluids?

BIOAVAILABILITY. After reading this section, answer the following.

41. Define "bioavailability."

42. What three things determine the bioavailability of a given nutrient?

43. Why is it recommended that nutrients be ingested from food rather than pills?

44. Why are fad diets not recommended for good nutrition?

TRANSPORT. As you read, fill in the blanks with the most appropriate term(s).

_____ 45. Carbohydrates may be absorbed into the blood in the form of ...

_____ 46. The only monosaccharide that is ultimately delivered to the body cells is ...

_____ 47. The generation of energy, stored in the form of ATP, occurs as a result of the process of ...

_____ 48. Glucose is first transported by the blood to the ...

_____ 49. Glucose molecules may be combined to form ... and stored in the liver.

_____ 50. Triglycerides, phospholipids, and cholesterol are reduced in size by ... in the small intestine.

180
200
A
active transport
B12
bile salts
C
cellular respiration
chyle
excreted
glycogen
glucose
HDLs
hypervitaminosis
lacteals
LDLs
lipoproteins
liver
micelles
monosaccharides
protein
stomach

_____ 51. Fat subunits, cholesterol, and phospholipids combine with bile salts to form small water-soluble ... which take the lipids to the epithelium.

_____ 52. Fat subunits pass through the small intestine and enter ... within the villi.

_____ 53. The fat-laden lymph is called ...

_____ 54. In the bloodstream, fats combine with proteins to form ... such as HDLs, LDLs, and VLDLs.

_____ 55. "Good cholesterol", or ... actually reduce the amount of fat in the bloodstream.

_____ 56. "Bad cholesterol", or ... are usually deposited into adipose cells.

_____ 57. Researchers recommend for people under 30 that blood cholesterol should not exceed ...

_____ 58. Protein digestion begins in the ... and continues in the small intestine.

_____ 59. Amino acids are transported across the intestinal epithelium by ... and enter the bloodstream.

_____ 60. Vitamin B12 must accompany ... to be absorbed through the intestinal epithelium.

180 mg/dl
200 g/dl
A
active transport
B12
bile salts
C
cellular respiration
chyle
excreted
glycogen
glucose
HDLs
hypervitaminosis
intrinsic factor
lacteals
LDLs
lipoproteins
liver
micelles
monosaccharides
protein
stomach

_____ 61. Though some amino acids may be metabolized in the liver or used to make plasma proteins, most of the amino acids are carried to body cells for the purpose of ... synthesis.

_____ 62. Water-soluble vitamins such as the B-complex vitamins and vitamin ..., are absorbed through the intestinal wall and transported by the bloodstream to the liver.

_____ 63. If the body does not need water-soluble vitamins, they are ...

_____ 64. Fat-soluble vitamins in excess can result in a poisoning condition called ...

180 mg/dl
200 mg/dl
A
active transport
B12
bile salts
C
cellular respiration
chyle
excreted
glycogen
glucose
HDLs
hypervitaminosis
intrinsic factor
lacteals
LDLs
lipoproteins
liver
micelles
monosaccharides
protein
stomach

STUDY HINT: It may help you to remember that HDLs are basically "Healthy", while LDLs are associated with a "Lethal" dose of cholesterol in the body tissues.

METABOLISM. Work through this section paragraph by paragraph. Study the diagrams carefully as they illustrate and summarize processes without excess words. Answer the following by selecting the choice which best completes each statement.

_____ 65. The collective process of building larger molecules from smaller ones and breaking down complex structures into simpler ones is
   a. metabolism
   b. anabolism
   c. catabolism
   d. cellular respiration

_____ 66. The digestion of food into its basic subunits for absorption is an example of
   a. an anabolic reaction
   b. a catabolic reaction
   c. a respiratory reaction

_____ 67. The breakdown of glucose, fatty acids, and amino acids within cells to release energy is
   a. cellular respiration
   b. catabolism
   c. anabolism
   d. both a and b are correct
   e. both a and c are correct

_____ 68. Oxidation can be best described as
   a. the breaking of the chemical bonds of nutrient molecules in the presence of enzymes
   b. the formation of energy-rich molecules that contain oxygen
   c. the catabolism of nutrient molecules when exposed to oxygen
   d. mechanical respiration

_____ 69. The series of biochemical reactions in the cell which results in the breakdown of glucose and the production of two ATP molecules is called
   a. cellular respiration
   b. glycolysis
   c. oxidation
   d. glycogenesis

_____ 70. When glycolysis occurs during strenuous exercise
   a. glycolysis essentially stops until oxygen is available
   b. the oxygen supply to the cell is reduced
   c. the pyruvic acid and energy produced in glycolysis will be processed by anaerobic respiration
   d. additional ATPs will be produced by aerobic respiration

_____ 71. Anaerobic respiration
   a. is a catabolic process
   b. is the continued breakdown of glucose molecules in the absence of oxygen
   c. results in the production of lactic acid and some ATP
   d. All of the above are true of anaerobic respiration

_____ 72. Aerobic respiration
   a. happens in the presence of oxygen
   b. results in 36 or 38 molecules of ATP for the breakdown of one glucose molecule
   c. results in 6 molecules of carbon dioxide which eventually leave the body in the exhalation from the lungs
   d. All of the above are true of aerobic respiration.

_____73. A cyclic series of reactions in which a citric acid molecule goes through a series of oxidation-reduction reactions is
   a. the electron transport chain
   b. the Krebs cycle
   c. cellular respiration
   d. none of the above

_____74. The breakdown of fatty acids
   a. involves the removal of two carbon atoms from the fatty acid chain to form acetyl CoA
   b. involves the production of ketone bodies in the presence of oxygen
   c. results in the formation of glycogen in the liver
   d. is known as ketosis and can be identified by the smell of acetone on the breath

_____75. Ketone bodies are most closely associated with
   a. fat metabolism
   b. protein metabolism
   c. carbohydrate metabolism
   d. anaerobic respiration

_____76. Given a choice between a ketone body and a glucose molecule as an energy source, a cell will use
   a. a ketone body
   b. a glucose molecule
   c. either one with equal frequency

_____77. Which of the following is true of protein synthesis?
   a. Amino acids from broken down proteins are excreted in urine.
   b. The amine group is removed, leaving ammonia and keto acid.
   c. The ammonia from amino acid breakdown is used in the electron transport chain to generate ATP.
   d. The goal of protein synthesis is the production of urine.

METABOLIC RATE AND BODY TEMPERATURE. Read this section aloud. Read as if you were trying to interest others in your research topic. Then briefly answer the following questions.

78. Define "metabolic rate."

79. You are a nurse. A patient has been told to report to you to determine his BMR. He is a little nervous about a BMR and asks what it is. What will you tell him?

80. What is the TMR?

81. "The oxidation of foods we eat is not completely efficient." Explain why that statement is true.

82. Why do hot, humid climates seem so much warmer than hot, dry climates?

83. How does the preoptic area of the hypothalamus affect homeostasis?

CLINICAL TERMS OF NUTRITION. Match these terms with their descriptions. Terms may be used more than once or not at all. When you have completed this exercise and checked your accuracy, review by supplying the description for each of the terms while covering the descriptions.

_____ 84.	caused by excessive body temperature (over 106 F)
_____ 85.	high levels of glucose in the blood
_____ 86.	genetic metabolic disease in which the amino acid phenylalanine cannot be metabolized properly
_____ 87.	overweight
_____ 88.	low levels of glucose in the blood
_____ 89.	white blood cells produce pytogens which stimulate the release of prostaglandins which cause the hypothalamus to raise body temperature
_____ 90.	may follow a high-glucose meal
_____ 91.	associated with mental retardation

fever
heat stroke
hyperglycemia
hypoglycemia
obesity
phenylketonuria

LABELS AND LISTS

1. List six main nutrients which are usually ingested into the body.

    a.   b.

    c.   d.

    e.   f.

2. List four fat-soluble vitamins.

    a.   b.

    c.   d.

3. List nine water-soluble vitamins. (Consult Table 17-2)

    a.   b.

    c.   d.

    e.   f.

    g.   h.

    i.

4. List seven major minerals required for good health.

    a.   b.

    c.   d.

    e.   f.

    g.

5. List three factors that determine the amount of a given nutrient the body can use.

    a.   b.

    c.

6. What five factors may cause an increase in the metabolic rate?

   a.   b.

   c.   d.

   e.

7. What four factors can decrease metabolic rate?

   a.   b.

   c.   d.

8. List and define four primary routes of heat loss for the body.

   a.

   b.

   c.

   d.

9. List three ways that the body can achieve a higher internal body temperature.

   a.   b.

   c.

ADDITIONAL STUDY

Read the Chapter Summary (pages 527 - 528). Write out the definitions of all the KEY TERMS (page 528).

Review the illustrations in your textbook and understand the answers to the questions associated with each one. The answers are on pages 529.

Write out the answers to the "CONCEPTS CHECKS" and "QUESTIONS FOR REVIEW" in your text.

The day before an exam over this chapter, read the "Learning Objectives", page 508, and review any of the sections which you think will cause you a problem.

CHAPTER 18: THE URINARY SYSTEM

KIDNEY STRUCTURES. As you read, refer to the figures of kidney structure with intense interest as if an aunt had asked you to accompany her to the doctor. Become familiar with the subject so that you can carry on an intelligent conversation with the doctor and explain everything to your aunt. When you have studied, fill in these blanks.

_____ 1. position of the kidneys	1,000,000
_____ 2. responsible for the lower position of the right kidney	1200 ml/minute
	adipose capsule
	fenestrae
_____ 3. outer connective tissue anchoring kidneys to	**glomerulus**
	juxtaglomerular apparatus
	liver
_____ 4. connective tissue providing kidney with cushioning layer	**podocytes**
	renal fascia
	renal corpuscle
	renal capsule
_____ 5. arteries carrying blood to the kidneys	**renal arteries**
	retroperitoneal
	renal veins

_____ 6. volume of blood that enters kidneys

_____ 7. approximate number of nephrons in each kidney

_____ 8. bulb-like end of a nephron

_____ 9. literally, little ball of yarn

_____ 10. literally, window

_____ 11. literally, foot cell

_____ 12. part of a nephron where the distal convoluted tubule contacts the afferent arteriole

345

KIDNEY FUNCTIONS. Perk up your interest. This bit of physiology is some of the most fascinating in the whole body! Mark the true statements "true," and rewrite the false statements to make them true.

_____ 13. About *100 gallons* of fluid is pushed through the kidneys daily.

_____ 14. *The majority* of the fluid that passes through the kidneys is excreted as urine.

_____ 15. The three processes that are involved in urine formation are filtration, reabsorption, and *secretion*.

_____ 16. *Glomerular filtration* is the movement of blood plasma across the filtration membrane of the renal corpuscle.

_____ 17. Proteins *readily* filter out of the blood across the filtration membrane of the renal corpuscle.

_____ 18. The fluid and dissolved substances that go through the filtration membrane are called the *filtrate*.

_____ 19. NFP, or *Nephron Filtration Pressure*, pushes large quantities of fluid and small molecules through the pores of the glomerular wall, basement membrane, and between the processes of the podocytes.

_____ 20. When the filtration pressure decreases, the volume of the filtrate *increases*, resulting in less urine formed.

_____ 21. Filtrate and urine composition are *essentially* the same.

_____22. A difference between plasma and filtrate is the presence or absence of *proteins*.

_____23. Compared to blood plasma, urine contains *more* ions, urea, and uric acid.

_____24. Reabsorption *always* involves the movement of substances from the renal tubule into the bloodstream.

_____25. Water and solute reabsorption mainly takes place in the *distal* convoluted tubule.

_____26. In secretion, substances move from the blood *into the renal tubule*.

_____27. Secretion mainly occurs in the *glomerulus* of the nephron.

_____28. Two substances removed from the filtrate by secretion are *urea and uric acid*.

_____29. The level of urea in the blood is related to the amount of *protein in the diet because it is a by-product of amino acid metabolism.*

_____30. Uric acid results from the metabolism of *proteins*.

_____31. The abrupt stoppage of kidney function is known as *chronic renal failure*.

_____32. Total kidney failure can cause death *within 12 days*, but can be treated by dialysis.

REGULATION. Fill in the blanks with the terms which best completes each statement concerning kidney regulation.

_____ 33. The enzyme released by juxtaglomerular cells which eventually causes more filtrate to be produced is ...

_____ 34. Considering the sequence of events which begin as the filtration rate declines. Place the following in order: angiotensin I, increase in filtration pressure, aldosterone, renin, angiotensin II, increased blood pressure, increased blood volume.
_____
_____
_____
_____
_____
_____

_____ 35. ... is the hormone which regulates the rate of active transport in the nephron.

_____ 36. ... is the hormone which regulates the amount of water that is reabsorbed by the distal convoluted tubules (more hormone, more water reabsorbed).

_____ 37. ... is the hormone released by cells in the heart which causes blood volume and thus blood pressure to be reduced.

_____ 38. The test for the concentration of various chemicals in a urine sample is a ...

_____ 39. Considering the urinalysis chart in the text, the substance in the urine which indicates possible urinary tract infections is ...

_____ 40. Considering the urinalysis chart, ... is not expected in normal urine and its presence could indicate diabetes mellitus or pituitary problems.

FACTORS CONTROLLING URINE CONCENTRATION AND VOLUME. Fill in the following chart.

FACTOR	SOURCE	EFFECT
renin	41.	ultimately leads to an increase in blood pressure and a more concentrated urine
42.	adrenal gland	43.
44.	posterior pituitary	increased water reabsorption, leading to more concentrated urine
45.	right atrium	decreased water and solute reabsorption
sympathetic impulses	46.	47.

MAINTENANCE OF BODY FLUIDS AND pH. Answer the following in brief, informative sentences.

48. In order for the body to maintain homeostasis, what must be the relationship between intake of substances and their removal?

49. What organ of the body has the greatest immediate effect on fluid balance?

50. What are the acceptable pH margins between which the body must be maintained?

51. What is acidosis, and how is it dangerous to the body?

52. What is alkalosis?

53. What is a buffer?

54. How does the respiratory system help maintain a balanced pH level in the blood?

55. What is the response of the kidneys to a drop in pH?

56. What is the response of the kidneys to a pH rise?

57. What is respiratory acidosis?

58. What is metabolic acidosis?

59. What is respiratory alkalosis? How is it related to hyperventilation?

60. What is metabolic alkalosis?

URETERS. After you read the short section describing the ureters and study their structure on the diagram, answer the following. Select the choice which best completes each statement.

_____61. The tubular organ that carries urine from the kidney to the urinary bladder is the
   a. urethra
   b. ureter

_____62. The position of the ureters in the body can be described as
   a. juxtapositioned
   b. antiperitoneal
   c. retroperitoneal
   d. anterior

_____63. The ureters are protected from the corrosive effect of urine by
   a. star cells
   b. rapid peristaltic waves
   c. mucus
   d. buffers

_____ 64. What prevents urine from backing up from the urinary bladder to the ureters?
   a. a sphincter muscle
   b. a flap of mucous membrane
   c. a fibroid filter
   d. fibrous connective tissue in a semilunar shape

URINARY BLADDER. Match the following terms with the statements which best describe them.

_____ 65. hollow, muscular organ responsible for urine storage on a temporary basis

_____ 66. average capacity of the urinary bladder

_____ 67. average maximum capacity of the urinary bladder

_____ 68. number of openings into the urinary bladder

_____ 69. region of the bladder that is a frequent site of urinary tract infections

_____ 70. innermost layer of the bladder

_____ 71. the thick layer of smooth muscle in the wall of the bladder active in urination

1 pint
2+ pints
3
detrusor
mucous membrane
trigone
urinary bladder

URETHRA. While you read about the urethra, fill in the blanks for the following statements. Thus, this section of study work will act as a pacer for your reading and will help you concentrate.

_____ 72. The ... is a muscular tube that drains urine from the floor of the urinary bladder and transports it out of the body.

_____ 73. The ... sphincter is located at the junction between the bladder and the urethra.

_____ 74. The internal urethral ... is involuntary and keeps urine from entering the urethra while it is being stored in the bladder.

_____ 75. Which sphincter is under voluntary control?

_____ 76. In which gender is the urethra longer?

MICTURITION Read about micturition before answering the following. Read with enough interest and enthusiasm for the topic that you could intelligently and accurately describe the process.

77. What are three medical terms for eliminating urine from the body.

78. What causes an urge to void?

79. If the time is not appropriate for voiding, how may urine flow be stopped?

HOMEOSTASIS. This portion of text is very practical. Read the section aloud. Then consider the situations that follow. Select the appropriate sequence of body responses that would maintain homeostasis. Place the letters of the sequence in the blanks. More importantly, be able to verbally list these body responses without assistance.

80. Jeremy plays a tennis tournament on a summer afternoon which results in one sweaty boy (water and salt loss). How does his body adjust to maintain homeostasis?

_____ _____ _____ _____

81. Elizabeth is a novice gourmet cook and has prepared pizza with salted anchovies, ham garnishes, and a salty pickle salad on the side. How do her guests' bodies maintain homeostasis in response to this heavy salt intake?

_____ _____ _____ _____ _____

a. ADH, aldosterone, and renin levels increased
b. more concentrated salt is secreted in urine
c. blood pressure remains relatively constant
d. lowered urine output
e. kidneys reabsorb more water and salt
f. ADH level is increased
g. kidneys reabsorb more H_2O

CLINICAL TERMS OF THE URINARY SYSTEM. Match these terms with their descriptions. Place the correctly spelled term in the blanks, and be sure to pronounce the words aloud to aid your memory.

_____ 82. inflammation of the urinary bladder

_____ 83. increased production of urine, perhaps due to a diuretic

_____ 84. inflammation of the glomeruli

_____ 85. absence of urine

_____ 86. kidney stones due to high salt concentration in urine

_____ 87. pain during urination

anuria
calculi
congenital polycystic kidney disease
cystitis
diuresis
dysuria
enuresis
glomerulonephritis
hematuria
oliguria
polyuria
pyelonephritis
uremia
urethritis
urinary tract tumors

_____ 88. uncontrolled urination
_____ 89. blood in the urine
_____ 90. inflammation of the urethra
_____ 91. toxic levels of urea in the blood, perhaps as a result of kidney failure
_____ 92. inflammation of renal pelvis, often caused by bacteria
_____ 93. excessive output of urine
_____ 94. reduced output of urine
_____ 95. abnormal development of renal tubules and collecting ducts

anuria
calculi
congenital polycystic kidney disease
cystitis
diuresis
dysuria
enuresis
glomerulonephritis
hematuria
oliguria
polyuria
pyelonephritis
uremia
urethritis
urinary tract tumors

LABELS AND LISTS

1. List three layers that surround the kidney.

    a.   b.

    c.

2. List the vessels through which blood flows as it travels through the nephron. Rearrange the following: efferent arteriole, glomerulus, afferent arteriole, peritubular capillaries

3. In what four ways are substances moved during reabsorption?

    a.   b.

    c.   d.

4. List five factors that control urine concentration and volume.

   a.   b.

   c.   d.

   e.

5. List three ways that pH in body fluids is controlled.

   a.   b.

   c.

6. Label the organs of the urinary system.

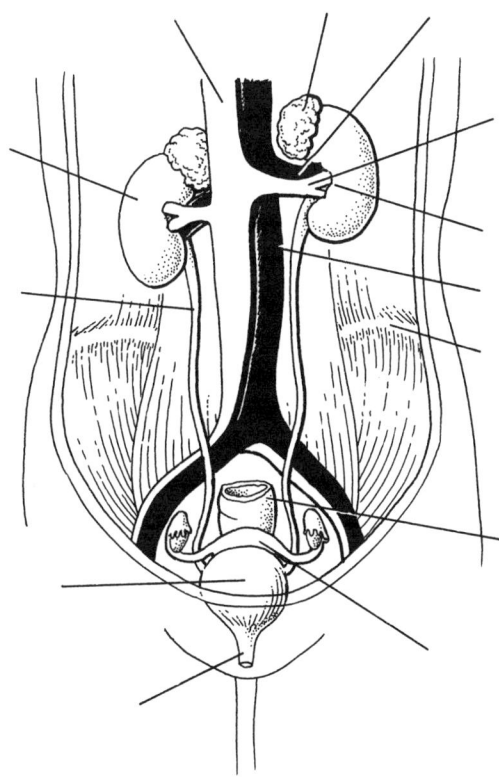

355

7. Label the kidney.

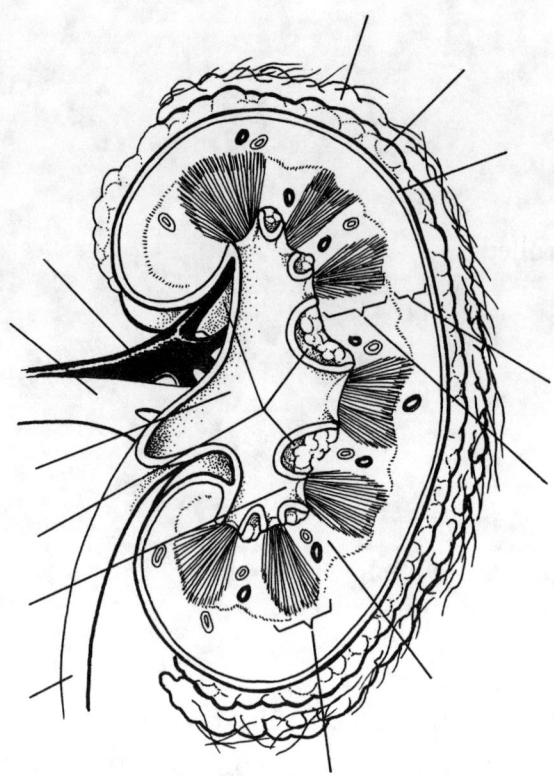

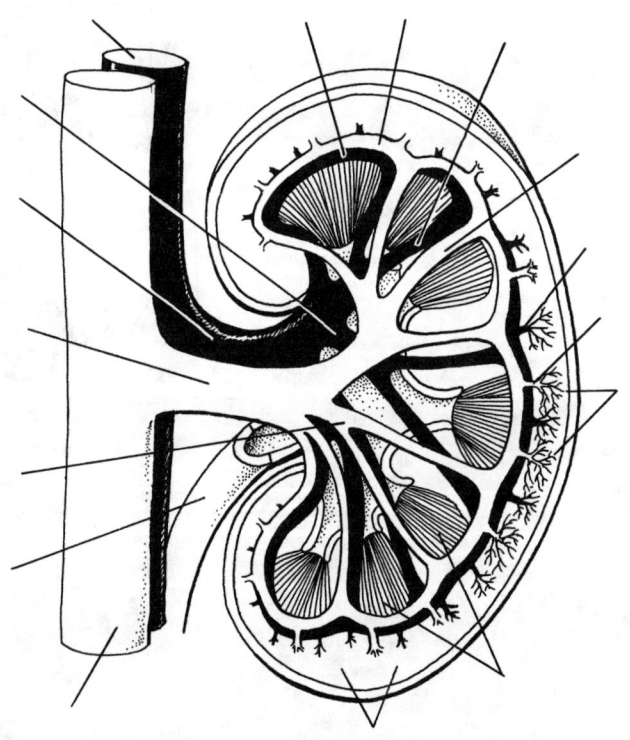

356

8. Label the nephron.

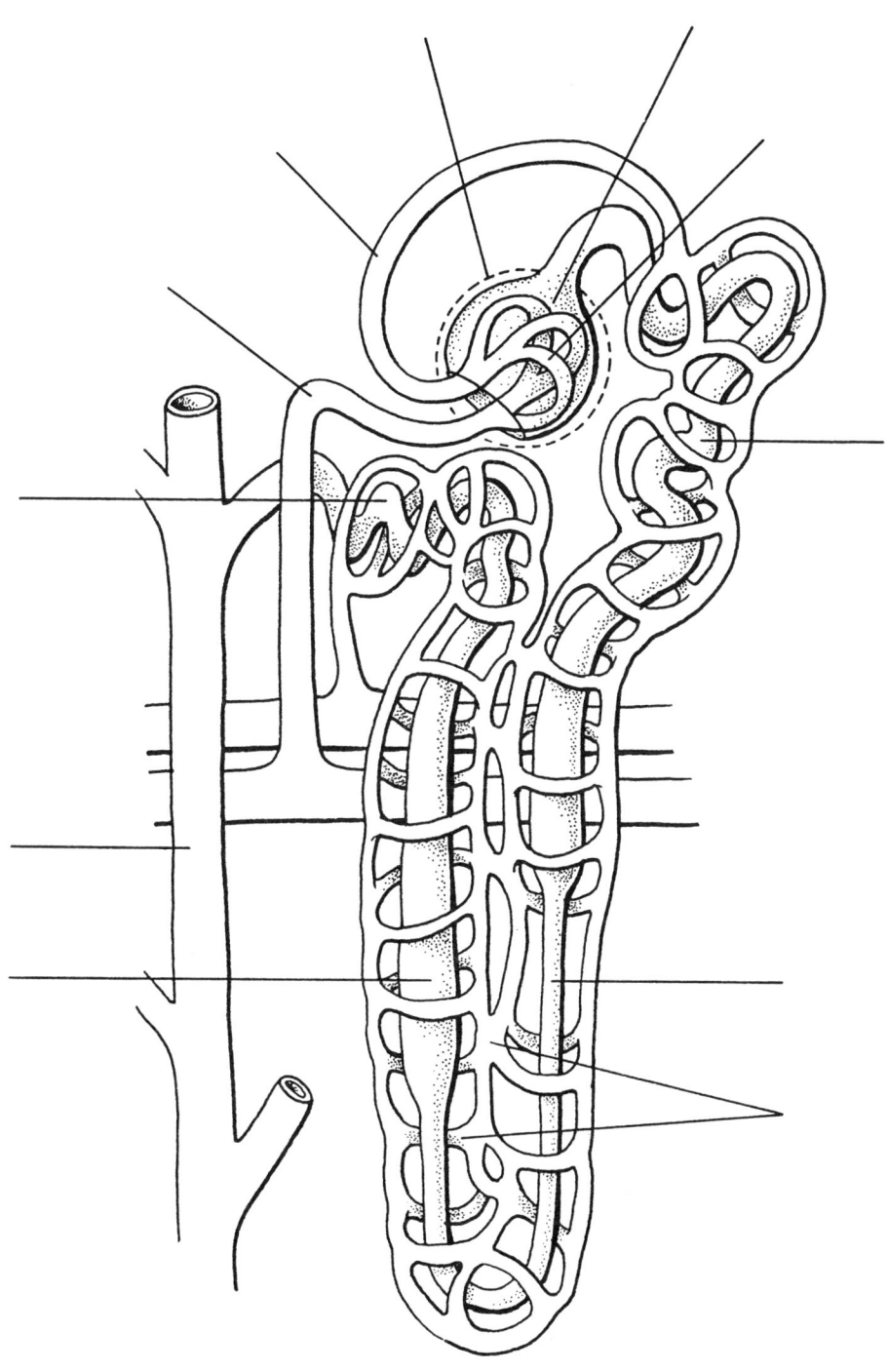

9. Label the structures.

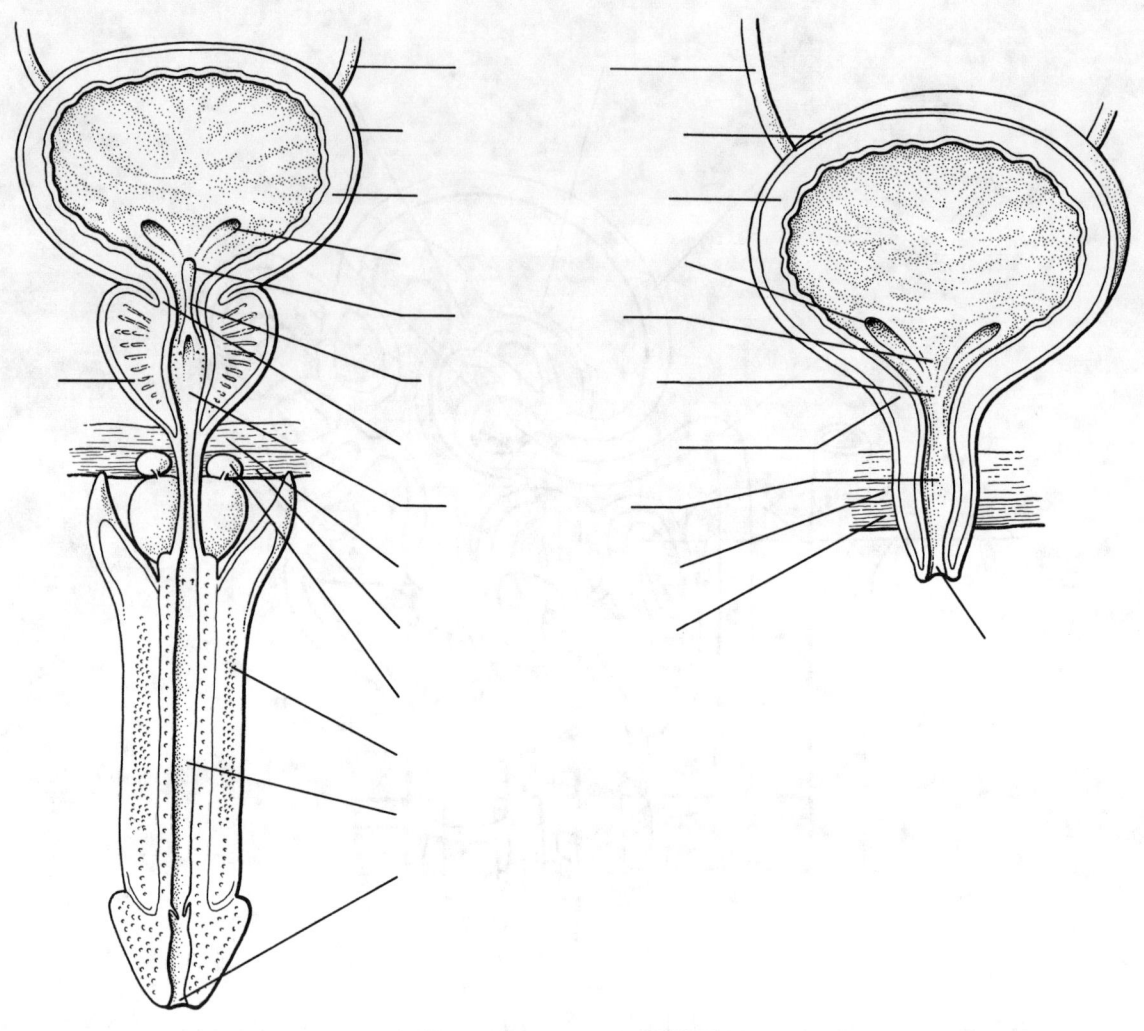

ADDITIONAL STUDY

Read the Chapter Summary (pages 551 - 552 and write the definitions of all the KEY TERMS (page 552).

Review the illustrations in your textbook and understand the answers to the questions associated with each one. The answers are on pages 553 - 554.

Having studied this chapter, close your book, put away your notes, and test yourself by **writing** the answers to the "Learning Objectives", page 530, and review any of the sections which you think will cause you a problem on future exams

CHAPTER 19: THE REPRODUCTIVE SYSTEM

CONTENT MASTERY

ORGANS OF MALE REPRODUCTION, TESTES. When you have worked through this section. Test yourself once more by defining each term without any prompting.

_____ 1. primary organs of the male reproductive system

_____ 2. male gonads

_____ 3. male sex hormone

_____ 4. male gametes

_____ 5. pouch holding the testes

_____ 6. literally, eggshell

_____ 7. literally, tiny tube to carry seed

_____ 8. cells of Leydig

_____ 9. cells that secrete testosterone

_____ 10. cells in the germinal epithelium of the seminiferous tubules that undergo spermatogenesis

_____ 11. the production of sperm cells

_____ 12. two types of cells that can be produced by division of spermatogonia

germ cells
germinal epithelium
interstitial cells
primary spermatocytes
scrotum
seminiferous tubules
spermatid
sperm
spermatogenesis
spermatogonia
sterility
testosterone
testes

STUDY TIP: Review meiosis in Chapter Three

_____ 13. before a spermatozoan developed a head and flagellum, it was this type of cell

**germ cells
germinal epithelium
interstitial cells
primary spermatocytes
scrotum
seminiferous tubules
spermatid
sperm
spermatogenesis
spermatogonia
sterility
testosterone
testes**

_____ 14. condition of insufficient sperm cells needed to fertilize an egg

DUCTS. After reading, fill in the blanks. Pay attention to spelling.

_____ 15. The first duct through which sperm travel after leaving the testes is the ...

_____ 16. The length of this duct is about ... meters long.

_____ 17. Sperm move through the epididymis by way of ... contractions.

_____ 18. Mature sperm move by way of ...

_____ 19. Another name for the vas deferens is the ...

_____ 20. The vas deferens enters the abdomen through the ...

_____ 21. The ductus deferens empties into the ... duct which carries sperm through the prostate gland.

_____ 22. The ... cord consists of the ductus deferens, the testicular artery, veins, lymph vessels, and nerve, the cremaster muscle, and connective tissue.

_____ 23. The cutting of the ductus deferens, or surgical sterilization, is also known as a ...

_____ 24. The tube which extends from the urinary bladder to the distal end of the penis is the ...

_____ 25. The opening of the urethra is called the ...

ACCESSORY GLANDS.

_____26. The collective function of the seminal vesicles, prostate gland, and the bulbourethral glands is
   a. to produce sperm cells
   b. to produce provide a liquid, nutritious medium for sperm cells
   c. to move sperm cells from the ductus deferens to the urethra
   d. all of the above

_____27. The mixture of sperm cells and fluids is called
   a. semen
   b. seminal fluid
   c. prostate secretion
   d. both a and b

_____28. The collection of glands located at the base of the urinary bladder is the
   a. vas deferens
   b. sperm glands
   c. seminal vesicles
   d. ejaculatory bladders

_____29. The number of prostate glands in the normal male is
   a. one
   b. two
   c. four
   d. hundreds

_____30. The result of hypertrophy of the prostate gland is
   a. enlarged urethral openings resulting in incontinence
   b. enlargement of the prostate resulting in a restricted urethra
   c. painful urination
   d. both b and c

_____31. The accessory organ(s) that produce(s) secretions rich in sugar is (are)
   a. the seminal vesicles
   b. prostate gland
   c. bulbourethral glands
   d. Cowper's glands

_____32. The gland(s) responsible for alkaline fluid to counteract vaginal acidity is (are)
   a. seminal vesicles
   b. prostate gland
   c. bulbourethral glands
   d. Cowper's glands

____33. An enlarged prostate is called
   a. hyperprostate benignity
   b. prostate hyperclimatica
   c. benign prostatic hyperplasia
   d. hypoprostate benignity

____34. Accessory organ(s) responsible for lubrication for sexual intercourse is/are
   a. seminal vesicles
   b. prostate gland
   c. bulbourethral glands

EXTERNAL GENITALIA. Match these terms with the descriptions given.

_____ 35. literally, bag

_____ 36. pouch of skin that hangs below the lower abdominal wall

_____ 37. muscle which helps to draw the testes closer to the abdominal wall in cold

_____ 38. external cylindrical organ

_____ 39. enlarged end of the penis

_____ 40. literally, bodies containing hollow areas

_____ 41. literally, body containing sponges

_____ 42. erectile tissue through which the urethra passes

_____ 43. surgical removal of the prepuce

circumcision
corpora cavernosa
corpus spongiosum
dartos
glans penis
penis
prepuce
scrotum

NOTE: "Two of the columns (of erectile tissue) form the dorsal and lateral portions of the penis and are called **corpora cavernosa**." (page 564) The dorsal portion of the penis is that portion which is dorsal during erection. Similarly, the corpus spongiosum is ventral during erection.

PHYSIOLOGY OF MALE REPRODUCTION. Fill in the following terms.

_____ 44. filling of tissues in the penis with blood

_____ 45. movement of sperm from the epididymis to the urethra

_____ 46. movement of semen through the urethra

_____ 47. male climax

_____ 48. three sources of hormones that influence male reproductive functions

_____ 49. hormone released by hypothalamus at puberty

_____ 50. hormone that causes anterior pituitary to secrete gonadotropins

_____ 51. two gonadotropins

_____ 52. collectively, the male sex hormones

_____ 53. the most abundant male sex hormone

_____ 54. the functional development of sexual characteristics

_____ 55. hormone that stimulates vocal cord enlargement

_____ 56. hormone that promotes development of interstitial cells in the testes

androgens
anterior pituitary gland
ejaculation
emission
erection
follicle stimulating hormone
gonadotropin-releasing hormone
hypothalamus
luteinizing hormone
orgasm
testosterone
testes

ORGANS OF FEMALE REPRODUCTION. As you read and study this section of material, pay close attention to the illustrations. When you are ready, fill in the blanks with the most appropriate term(s). Pay close attention to spelling.

_____ 57. In the female, the primary sex organs are the ...

_____ 58. Saclike structures in the cortex of the ovary are called...

_____ 59. Names for the female gamete. (two)

_____ 60. The maturation and release of oocytes on a monthly schedule is called ...

_____ 61. The oocytes that are present in the ovary at birth are called ...

_____ 62. Egg development is called ...

_____ 63. The structure of cells around a primary oocyte is a ...

_____ 64. Approximately ... primary oocytes are stimulated to begin meiosis each month.

_____ 65. A secondary oocyte is stimulated to undergo its second meiotic division by ...

_____ 66. The fertilized egg is known as a ...

_____ 67. The cavity within a primary follicle which forms between the oocyte and follicle cells is called the ...

_____ 68. The follicle cells of a secondary follicle produce the primary female sex hormone, ...

_____ 69. The mature, 10-day follicle is called the ... follicle

_____ 70. The expulsion of the Graafian follicle from the ovary is known as ...

_____ 71. literally, encircling crown

_____ 72. literally, yellow body

_____ 73. The outer covering of the expelled oocyte is the ...

_____ 74. Fertilization usually occurs in the ...

_____ 75. Cells of the ... secrete progesterone and estrogen.

_____ 76. If fertilization occurs, the hormones produced by the ... maintain pregnancy for three months until the placenta takes over.

FEMALE ACCESSORY ORGANS. Terms may be used more than once or not at all.

_____ 77. another term for uterine tube

_____ 78. structure on which fimbriae are found

_____ 79. literally, funnel

_____ 80. literally, fringes

_____ 81. site of menstruation

_____ 82. literally, neck

_____ 83. literally, sheath

_____ 84. three distinct layers of the uterus

_____

_____

_____ 85. inner lining of the uterus, part of which is sloughed off during menstruation

_____ 86. birth canal

_____ 87. glands that provide lubrication for the vagina

_____ 88. opening to the outside

cervix
endometrium
fallopian tube, or oviduct
fertilization
fimbriae
fornix
infundibulum
myometrium
serous
uterus
vaginal orifice
vagina
vestibular glands

EXTERNAL GENITALIA. Match these terms with their descriptions. Terms may be used more than once or not at all.

_____ 89. literally, presence of a covering

_____ 90. area over symphysis pubis covered with hair after puberty

_____ 91. female homologue to male scrotum

_____ 92. female homologue to the male penis

_____ 93. female homologues to male bulbourethral glands

_____ 94. the region between the upper border of the vestibule and the anus

_____ 95. cutting of the perineum to avoid tearing during childbirth

_____ 96. literally, prominent lips

_____ 97. literally, outer chamber

clitoris
episiotomy
labia majora
mons pubis
perineum
prepuce
vestibule
vestibular glands
vulva

MAMMARY GLANDS. Read about the mammary glands. Briefly answer these questions.

98. What causes the breasts to enlarge?

99. What are lobes, lobules, and alveolar glands?

PHYSIOLOGY OF FEMALE REPRODUCTION. The "hows" and "whys" of female reproductive physiology are fascinating. To work through this material, use the following reading pacer. The statements to complete are in the order presented in the text. Occasionally, statements are reworded for repetition to better enable you to retain the information.

_____ 100. When the influence of hormones and neural mechanisms are compared, ... have the greater influence on female reproduction.

_____ 101. Hormones control reproductive organ development, ..., menstruation, and maintain pregnancy.

_____ 102. Hormones control organ development, ovulation, ..., and maintain pregnancy.

_____ 103. Hormones control organ development, ovulation, menstruation, and maintain ...

_____ 104. Neural mechanisms govern the physiological responses to ...

_____ 105. The vulva and sensitive ... respond to stimulation by way of an involuntary reflex arc.

_____ 106. The parasympathetic impulses stimulate secretion of ... by the vestibular glands for lubrication.

_____ 107. The parasympathetic impulses stimulate secretions by the ... for lubrication.

_____ 108. The orgasm reflex involving peristaltic contractions of muscles in the perineum, vagina, uterus, and uterine tubes promote sexual stimulation and may aid the movement of ... through the female tract.

_____ 109. Hormones that influence the female reproductive system arise from the ..., the anterior pituitary, and the ovaries.

_____ 110. Hormones that influence the female reproductive system arise from the hypothalamus, the ..., and the ovaries.

_____ 111. Hormones that influence the female reproductive system arise from the hypothalamus, the anterior pituitary, and the ...

_____ 112. Throughout childhood, the ovaries grow slowly and continuously secrete the hormone...

_____ 113. Near puberty, the hypothalamus begins to release ...

_____ 114. GnRH is released by the ...

_____ 115. Once released, GnRH diffuses to the ... and causes it to secrete two hormones.

_____ 116. GnRH stimulates the anterior pituitary to secrete ... and ... which affect the ovaries.

_____ 117. FSH acts upon the primary follicles of the ... to cause them to mature.

_____ 118. The maturation of the follicles causes them to produce increasing levels of ... and ...

_____ 119. LH also promotes the secretion of ... and ...

_____ 120. The effects of increased estrogen and progesterone levels brings about ... and puberty.

_____ 121. At puberty, estrogen ... the female body by causing the enlargement of the vagina, uterus, uterine tubes, and external genitals.

_____ 122. Estrogen also promotes the accumulation of ... in the breasts, thighs, and buttocks.

_____ 123. Ovaries also secrete the sex hormone, ... which works with estrogen to bring about uterine changes during the menstrual cycle.

_____ 124. Progesterone is produced in the ... in the ovaries and during pregnancy by the placenta.

_____ 125. The ... or uterine cycle is a series of cyclic changes that occur in sexually mature females until about the age of 50.

_____ 126. Another term for menstrual bleeding is ...

_____ 127. Menstrual bleeding is a mild hemorrhage caused by sloughing off of part of the ...

_____ 128. According to Figure 19-14, the regeneration of the endometrium is due to rising levels of ...

_____ 129. The period of growth of the endometrium during the first half of the menstrual cycle is called the ... phase.

_____ 130. As estrogen levels in the blood increase during the first 14 days, the endometrium gradually thickens; the hypothalamus produces more ...

_____ 131. GnRH stimulates more ... and ... from the pituitary.

_____ 132. The rise in LH stimulates ... about day 14 of the menstrual cycle.

_____ 133. After ovulation, the follicle cells in the ovaries are converted to a ... which begins to secrete progesterone and small amounts of estrogen.

_____ 134. After ovulation, the corpus luteum secretes ... and small amounts of estrogen.

_____ 135. Progesterone levels thus begin to ... and inhibits the LH and FSH secretion.

_____ 136. Progesterone also promotes development of the ... by stimulating cell growth and the secretion of nourishing fluid.

_____ 137. The period of time between ovulation and the next menses is called the ... phase.

_____ 138. After about 7 to 8 days after ovulation, the ... is fully prepared to receive an early embryo.

_____ 139. If the oocyte is unfertilized, the ... begins to degenerate, progesterone and estrogen levels drop, and the outer layer of the endometrium flows out of the uterus.

_____ 140. Another term for menstrual cramps is ...

_____ 141. The pain associated with menstrual cramps is a result of an over production of ...

_____ 142. PMS, or ..., is thought to be caused by imbalances in hormone levels during the latter half of the menstrual cycle.

_____ 143. The period in life (about age 45 or 50) during which the menstrual cycles are less regular is referred to as female ...

_____ 144. The cessation of the menstrual cycle is a time in a woman's life called ...

_____ 145. The prevention of pregnancy is termed ...

_____ 146. Male sterilization called ... and involves cutting the ductus deferens.

CLINICAL TERMS OF THE REPRODUCTIVE SYSTEM.

_____ 147. caused by the bacterium *Neisseria gonorrhoeae*

_____ 148. caused by a virus, Herpes simplex genitalia

_____ 149. uterectomy

_____ 150. infection of the uterine tubes and adjacent tissues

_____ 151. caused by a toxin from the bacterium *Staphylococcus aureus*

_____ 152. inflammation of the vaginal mucous membrane

_____ 153. absence of menses

_____ 154. most common malignant tumor in women

_____ 155. severe pain during menstruation, sterility, and painful intercourse

_____ 156. malignancy of the prostate gland

_____ 157. cancer arising from epithelial cells of cervix

_____ 158. failure of the testes to descend into the scrotum

amenorrhea
breast cancer
cancer of the prostate
cervical cancer
cryptorchidism
dilation and curettage
endometriosis
fibroadenoma
fibrocystic disease
gonorrhea
herpes
hysterectomy
ovarian tumors and cysts
pelvic inflammatory disease
syphilis
toxic shock syndrome
trichomoniasis
vaginitis

_____ 159. surgical procedure in which the cervix is dilated, and the endometrium is scraped

_____ 160. abnormal growths of breast tissue like small pebbles beneath the skin

_____ 161. common benign breast tumor in young women

_____ 162. sexually transmitted disease caused by the bacterium *Treponema pallidum*

_____ 163. sexually transmitted disease caused by the protozoan *Trichomonas vaginalis*

_____ 164. usually caused by the fungus *Candida albicans*, the protozoan *Trichomonas vaginalis*, or bacterium *Gardnerella vaginalis*

_____ 165. surgical removal of the uterus

_____ 166. also known as D&C

_____ 167. Usually occurs in menstruating women who use high-absorbency tampons

_____ 168. if untreated, this sexually transmitted disease can ultimately lead to destruction of the nervous system

amenorrhea
breast cancer
cancer of the prostate
cervical cancer
cryptorchidism
dilation and curettage
endometriosis
fibroadenoma
fibrocystic disease
gonorrhea
herpes
hysterectomy
ovarian tumors and cysts
pelvic inflammatory disease
syphilis
toxic shock syndrome
trichomoniasis
vaginitis

LABELS AND LISTS

1. List four different ducts through which sperm travel.

    a.	b.

    c.	d.

2. List seven components that form the spermatic cord.

    a.	b.

    c.	d.

    e.	f.

    g.

3. What three types of organs serve as accessory glands to the male reproductive tract?

    a.	b.

    c.

4. The hormones that influence male reproductive functions arise from what three sources?

    a.	b.

    c.

5. List five hormones that effect female reproduction.

    a.	b.

    c.	d.

    e.

6. List three types of organs which serve as accessory organs for the female reproductive system.

   a.   b.

   c.

7. Label the drawing.

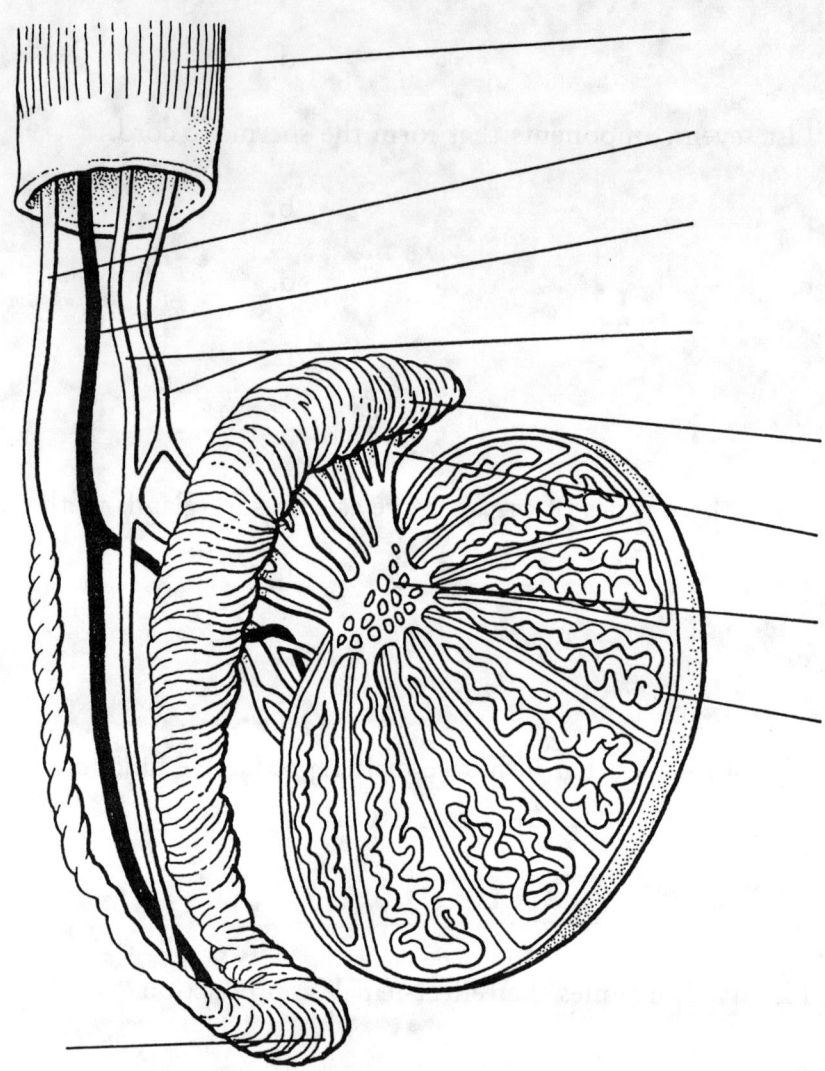

8. Label the parts of a sperm cell.

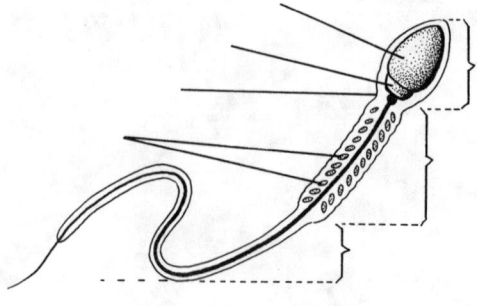

9. Label the female reproductive anatomy.

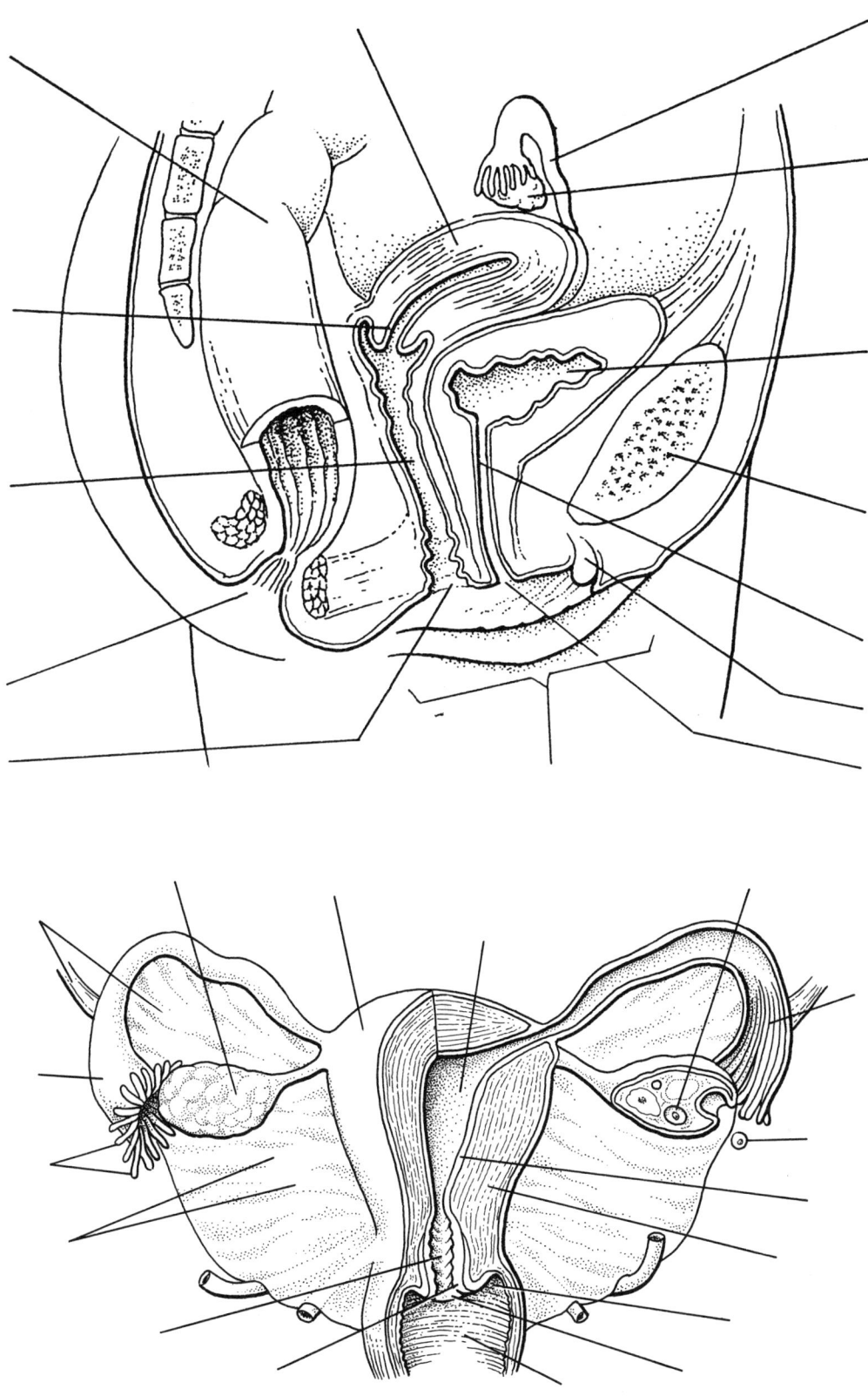

10. Label the figures.

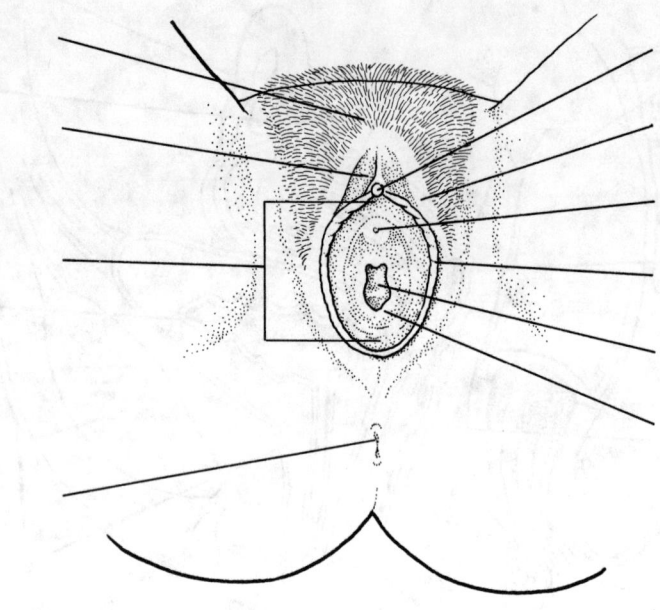

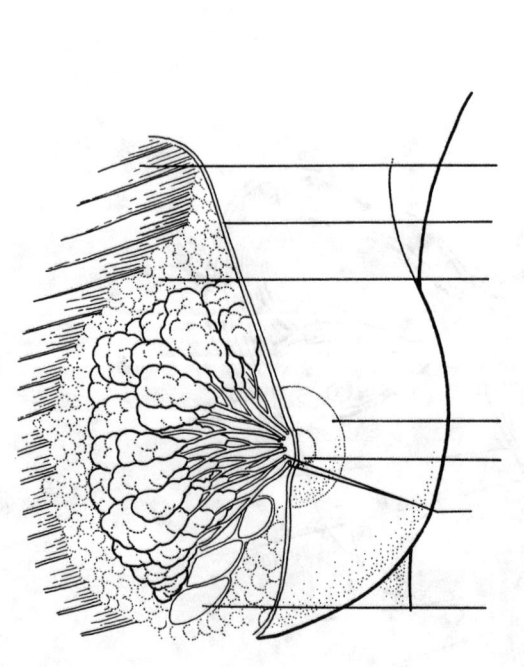
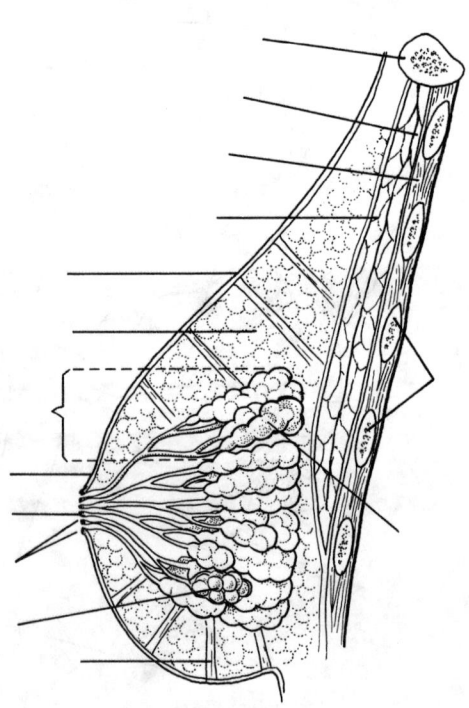

ADDITIONAL STUDY

After you have worked through the chapter using the study guide, read the Chapter Summary (pages 581 - 583). Write out the definitions of all the KEY TERMS (page 583). Then, review the figures in your textbook and understand the answers to the questions associated with each one. The answers are on pages 584 - 585.

Having studied this chapter, close your book, put away your notes, and test yourself by **writing** the answers to the "CONCEPTS CHECKS" and "QUESTIONS FOR REVIEW" in your text (pages 561, 563, 565, 566, 571, 572, 574, 577, 583 - 584). Writing the answers will force you to challenge yourself. If you can write the answers for yourself, you can probably write the answers for your professor.

The day before an exam over this chapter, read the "Learning Objectives", page 556, and review any of the sections which you think will cause you a problem.

REPRODUCTIVE SYSTEM
by Robert W. Bauman, Jr., Ph.D.

Across

3. disease caused by Treponema
4. surgical removal of the uterus
6. STD caused by virus
9. prevention of pregnancy
10. male accessory gland
11. mature follicle
13. cells which produce testosterone
15. erectile tissue in females
18. male accessory glands
21. foreskin
22. sex cells
23. male accessory glands
24. storage place for sperm cells
27. fluid and sperm
30. fringes
33. release of ovum
35. climax
37. lining of uterus
40. female sex organ
44. primary androgen
45. neck of uterus
46. gland which produces milk
47. muscle of spermatic cord

Down

1. induces ovulation
2. distal portion of penis
3. erectile body enclosing the urethra
5. male sex gland
7. primary female hormone
8. sex organs
9. erectile bodies which make up most of the penis
10. male organ of intercourse
12. pigmented portion of breast
14. funnel-shaped end of Fallopian tube
16. sac around testes
17. womb
19. cycle which results in menses
20. production of ova
25. promotes regeneration of endometrium
26. duct from epididymis
28. end of menstrual cycle
29. body which produces progesterone
31. nuclear division which halves the number of chromosomes
32. lips
34. tube which carries ova
36. STD caused by Neisseria
38. muscle of scrotum
39. birth canal
41. stimulates follicles to develop
42. nonfunctional cells from meiosis
43. mucosal barrier at vaginal orifice

- Crosswords Plus

378

CHAPTER 20: HUMAN DEVELOPMENT AND INHERITANCE

It is awe-inspiring to consider the development of a human being from a single fertilized cell to a tender, precious, newborn baby. As you have learned from studying the adult human body, even the apparently simple things of life involve phenomenal and intricate anatomical and physiological details. Such is the case in human development and inheritance. It is a good thing that complete awareness or understanding of the processes of development are not required for good health. However, your knowledge on this subject will enrich your appreciation of the human body.

CONTENT MASTERY

PRENATAL DEVELOPMENT. Carefully read the text material. When you have finished, scan back through and memorize the terms in bold print. Then, match these terms with their descriptions. Terms may be used more than once or not at all. If you would prefer the challenge, answer the questions without looking at the choices. This matching could thus becomes a review test rather than a workbook exercise. Grade yourself with a strict penalty for misspelled words!

_____ 1. a study of the changes that occur in prenatal development

_____ 2. refers to changes such as body growth, puberty, and senescence

_____ 3. the beginning of the prenatal period of development

_____ 4. period of prenatal development which occurs during the first 8 weeks

_____ 5. period of prenatal development which occurs during the last 38 weeks before birth

_____ 6. the union of an oocyte with a sperm cell

1 hour
24 hours
corona radiata
diploid (46)
embryology
embryonic period
fetal period
fertilization
haploid (23)
hyaluronidase
infertility
in vitro
postnatal development
prenatal development
zona pellucida
zygote

_____ 7. length of time for sperm travel within female reproductive tract

_____ 8. length of time during which oocyte is capable of being fertilized (after ovulation)

_____ 9. enzyme released from head of sperm to allow penetration of cells surrounding the oocyte

_____ 10. outer group of follicle cells through which sperm cell must travel to reach the oocyte

_____ 11. term describing chromosomal number in oocyte

_____ 12. term describing chromosomal number in sperm

_____ 13. term describing chromosomal number in zygote

_____ 14. fertilized cell, literally, presence of a yoke

_____ 15. most frequently caused by blocked uterine tubes and low sperm counts

_____ 16. type of fertilization which involves fertilization outside the body and subsequent artificial implantation

1 hour
24 hours
corona radiata
diploid (46)
embryology
embryonic period
fetal period
fertilization
haploid (23)
hyaluronidase
infertility
in vitro
postnatal development
prenatal development
zona pellucida
zygote

FIRST EIGHT WEEKS OF LIFE. As you read about the embryonic period of life, circle the answer in parenthesis which best completes each statement or question. Then, cover the choices, ask the question again, and answer as if it were a fill-in-the-blank question. In this way, take the time to learn each statement as you encounter it.

17. The zygote begins to divide by (mitosis/meiosis).

18. The process of division of a zygote is called (mitosis/cleavage).

19. During its first three days of development, the zygote is located in the (ovary/uterine tube/uterus).

20. After three days, the ball of cells resulting from cleavage of the zygote is called (blastula/morula/zygote).

21. The differentiation of cell types in the morula results in an outer layer of cells called the (zygote/trophoblast/trichoblast).

22. The Inner layer of cells is called the inner call mass and later becomes the (zygote/blastula/embryo).

23. Eventually, the intercellular spaces in the morula form a large cavity called the (trophoblast/blastocoele).

24. When the blastocyst cavity is formed, the developing organism is called a (morula/blastocyst).

25. The process of becoming "buried" in the endometrium is called (trophoblastocystulation/implantation).

26. Of the three germ layers, some of the cells of the (endoderm/ectoderm/mesoderm) eventually develop into the liver and gallbladder.

27. Of the three germ layers, some of the cells of the (endoderm/ectoderm/mesoderm) eventually develop into the bones.

28. Of the three germ layers, some of the cells of the (endoderm/ectoderm/mesoderm) eventually develop into the brain.

29. Implantation is complete in the third week of development when the trophoblast differentiates to form the (chorion/placenta/amnion/yolk sac).

30. A membrane that begins to develop during the third week which will eventually envelop the embryo is the (chorion/placenta/amnion).

31. The mesoderm forms during the process called (gastrulation/neurulation).

32. When the primitive streak of the embryo invaginates, the embryo is referred to as a (morula/gastrula/blastula).

33. The process of ectodermal growth in which the notochord, neural tube, and neural crest are formed is called (gastrulation/neurulation).

34. During the development of the body form (third to eighth weeks), the (foregut/hindgut/midgut) is a head cavity.

35. The combination of portions of the yolk sac and the allantois forms the (midgut/umbilical cord/embryonic disk).

36. The fluid which baths the embryo within a sac is the (chorionic/amniotic/allantoic) fluid.

37. The process of development of major organs which takes place during the first eight weeks of life is called (differentiation/organogenesis).

38. The heart of the embryo begins to beat about (21/201) days after fertilization.

39. By the end of the (fifth/seventh/fifteenth) week, most organs have formed to appoint of beginning their early functions.

40. The embryonic tissues produce finger-like projections called (chorionic villi/lacunae/microvilli) into the endometrium during the formation of the placenta.

41. The umbilical cord contains (one/two) umbilical artery(ies) and (one/two) umbilical vein(s).

42. At what point does the placenta take over the production of estrogen and progesterone? (months 1-3/months 4-9)

43. The virus that causes rubella (is/is not) capable of crossing the placental barrier.

44. *Toxoplasma gondii*, thalidomide, alcohol, and HIV virus (can/can not) cross the placental barrier to harm the developing embryo.

45. About eight weeks after fertilization, a human is called a (embryo/fetus).

46. The fine, soft hair covering the skin of the fetus is called (hair/fur/lanugo).

47. The fetus is said to be "full term" at (24/26/38) weeks.

PARTURITION. Read about the birth process, and define each of the following terms. Define them in full, knowledgeable sentences.

48. parturition -

49. Braxton-Hicks contractions -

50. oxytocin -

51. labor -

52. "water bag" rupture -

53. cervical dilation -

54. first stage of labor -

55. second stage of labor -

56. third stage of labor -

POSTNATAL DEVELOPMENT. Read each section carefully. Note the terms in bold print by writing your own definitions. If obvious questions come to mind as you read, write them (and their answers for future review). When you have worked through the section on your own using good study techniques, answer the following multiple choice questions as a review test.

_____57. Lactation can be best defined as
   a. the first year of life
   b. the production of milk by the mammary glands
   c. the months after parturition
   d. all of the above

_____58. Prolactin is a hormone which stimulates milk production. Since this hormone is present in the blood during pregnancy, why is milk not produced before the birth of the baby?
   a. The extra weight of the mother causes constriction of blood vessels in the mammary glands until after birth of the child.
   b. The placenta produces estrogen and progesterone which inhibits milk secretion.
   c. Actually, prolactin is not produced until labor.
   d. The loss of amniotic fluid reduces the blood pressure significantly and allows the hormone prolactin to be more effective in stimulating milk production.

_____59. During the first several days of milk production, there is a substance released by the mammary glands. This substance is called
   a. partuisterone
   b. lactose
   c. colostrum
   d. oxytocin

_____60. The hormone which is released from the posterior pituitary gland to stimulate milk "letdown" is
   a. progesterone
   b. oxytocin
   c. estrogen
   d. prolactin

_____61. A "neonate" is
   a. a newborn infant
   b. a prematurely born infant
   c. literally, the new presence of
   d. both a and c

_____ 62. Which of the following is true of the foramen ovale?
   a. Increased pressure of the left atrium causes a flap to close off the foramen ovale.
   b. Closure of the foramen ovale occurs with the first few breaths of the newborn.
   c. Foramen ovale literally means "oval shaped hole."
   d. All of the above are true.

_____ 63. Which of the following closes at or soon after birth? (There are multiple answers. Write as many as are accurate.)
   a. foramen ovale
   b. fossa ovalis
   c. ductus arteriosus
   d. ductus venosus

_____ 64. Place the stages of life in order.
   a. childhood, adulthood, adolescence, senescence
   b. senescence, childhood, adolescence, adulthood
   c. adolescence, childhood, senescence, adulthood
   d. childhood, adolescence, adulthood, senescence

_____ 65. Alzheimer's disease is
   a. a progressive brian disorder.
   b. a disease which generally afflicts persons 65 or older.
   c. a disease which causes gradual loss of memory and motor skills.
   d. all of the above.

GENETIC INHERITANCE. Carefully read "Genetic Inheritance" and "Chromosomes and Genes." While you read, memorize the bold terms. Then match the following with their descriptions.

_____ 66.	the study of how a variation of traits occurs among individuals	23
		46
		Aa
		aa
_____ 67.	number of chromosomes in every non-gamete body cell	**alleles**
		chromosomes
		genotype
_____ 68.	two chromosomes in a pair = ... chromosomes	**genetics**
		heterozygous
		homologous
_____ 69.	genes which carry the same traits (such as eye color) and occupy the same position on homologous chromosomes	**homozygous**
		phenotype

_____ 70. identical alleles for a trait

_____ 71. different alleles for a trait

_____ 72. a term that describes the actual genes a person has for a particular trait

_____ 73. a term that describes the appearance or physical expressions of the genes for a particular trait

_____ 74. if "A" represents normally pigmented skin, and "a" represents albinism, the designation for a heterozygous genotype

_____ 75. if "A" represents normally pigmented skin, and "a" represents albinism, the designation for a homozygous genotype

DOMINANT-RECESSIVE INHERITANCE. Carefully read this section of material aloud. Reading aloud will help focus your attention. Then answer the following.

76. What is the difference between a dominant gene and a recessive gene?

77. What is a Punnett square?

78. Consider Freckles (dominant gene "F") and absence of freckles (recessive gene "f"). Draw a Punnett square to illustrate the possible offspring when a heterozygous freckled man marries a woman who does not have freckles (just a few moles, which don't count in this question).

SEX-LINKED INHERITANCE. Read the material aloud. As you read terms that are highlighted in bold print, reread the definitions or descriptions. Then fill in the blanks with the best answers.

_____ 79. The chromosomes that are different in males and females are ...

_____ 80. The 22 pairs of chromosomes that are similar in both males and females are called ...

_____ 81. The sex chromosomes of a normal female consist of two ...-shaped chromosomes.

_____ 82. The sex chromosomes of a normal male consist of ... chromosomes.

_____ 83. The presence of the Y chromosome makes a person ... (gender).

_____ 84. The absence of a Y chromosome makes a person ... (gender)

_____ 85. Inherited traits that are determined by genes on the sex chromosomes are called ... traits.

_____ 86. Genes found on the X chromosome, but not on the Y, are called ...

_____ 87. Genes found on the Y chromosome are very rare, and are called ...

_____ 88. Can females inherit a Y-linked trait? (yes/no)

GENETIC SCREENING. Answer the following.

89. Define amniocentesis.

90. What blood tests are usually a part of routine genetic screening in maternity hospitals?

CLINICAL TERMS OF HUMAN DEVELOPMENT. Match these terms with their descriptions. Terms must be spelled correctly and may be used more than once.

_____ 91. spontaneous abortion

_____ 92. loss of mental or physical ability due to old age

_____ 93. spontaneous or deliberate termination of pregnancy

_____ 94. XXY

_____ 95. accumulation of bilirubin in the blood of a newborn infant

_____ 96. development of an embryo outside of its normal location as in peritoneal cavity

abortion
amniocentesis
cesarean section
congenital defects
ectopic pregnancy
fetal jaundice
hydramnios
hydatid mole
Klinefelter's syndrome
meconium
miscarriage
placenta previa
preeclampsia
senility
septal cardiac defects
tubal pregnancy
Turner's syndrome

_____ 97. removal of infant through a surgical incision

_____ 98. diagnostic procedure in which a sample of amniotic fluid is withdrawn from the amniotic cavity

_____ 99. uterine tumor that originates from placental tissue

_____ 100. fetal discharge that usually includes blood, serous fluid, and fluid wastes from the fetus

_____ 101. a genetic defect in females in which there is only one X chromosome per cell resulting in dwarfism among other anomalies

_____ 102. attachment of the placenta to the uterine wall near the cervical canal

_____ 103. a pregnancy-related syndrome characterized by sudden high blood pressure, protein in the urine, and edema throughout body tissues

_____ 104. type of ectopic pregnancy in which the fetus develops within the uterine tube

_____ 105. a variety of defective conditions in the newborn that are due to errors made during embryonic or fetal development

abortion
amniocentesis
cesarean section
congenital defects
ectopic pregnancy
fetal jaundice
hydramnios
hydatid mole
Klinefelter's syndrome
meconium
miscarriage
placenta previa
preeclampsia
senility
septal cardiac defects
tubal pregnancy
Turner's syndrome

_____ 106. excessive amount of amniotic fluid in the amniotic cavity

_____ 107. results in cyanotic congenital heart disease

abortion
amniocentesis
cesarean section
congenital defects
ectopic pregnancy
fetal jaundice
hydramnios
hydatid mole
Klinefelter's syndrome
meconium
miscarriage
placenta previa
preeclampsia
senility
septal cardiac defects
tubal pregnancy
Turner's syndrome

LABELS AND LISTS

1. List the stages of development from fertilization to parturition by placing the following terms in the correct sequence: **neonate, embryo, morula, zygote, neurula, blastocyst, fetus, gastrula**

2. List three factors which form the placental blood barrier to prevent the actual mixing of embryonic and maternal blood.

    a.   b.

    c.

3. List five structures that are functional before birth that close or disintegrate at birth or soon thereafter. Tell what structure they become, if the text discusses it.

    a.

    b.

    c.

    d.

    e.

ADDITIONAL STUDY

Read the Chapter Summary (pages 611 - 612). Write out the definitions of all the KEY TERMS (page 612).

Review the illustrations in your textbook and understand the answers to the questions associated with each one. The answers are on pages 613 - 614.

Having studied this chapter, close your book, put away your notes, and test yourself by **writing** the answers to the "CONCEPTS CHECKS" and "QUESTIONS FOR REVIEW" in your text.

The day before an exam over this chapter, read the "Learning Objectives", page 596, and review any of the sections which you think will cause you a problem.

APPENDIX

ANSWERS - CHAPTER 1

DIVISIONS OF STUDY
1. physiology
2. anatomy
3. regional anatomy
4. anatomy
5. gross anatomy
6. histology
7. microanatomy
8. physiology
9. systemic anatomy
10. anatomy
11. physiology

BASIC TERMINOLOGY
12. superior/cranial
13. inferior/caudal
14. anterior/ventral
15. posterior/dorsal
16. lateral
17. superficial/external
18. deep/internal
19. proximal
20. distal
21. superior
22. medial & inferior
23. anterior
24. posterior/dorsal
25. medial
26. lateral
27. superficial/external
28. deep/internal
29. proximal
30. distal

STRUCTURAL LEVELS OF ORGANIZATION
31. macromolecule
32. cells
33. cells
34. tissue
35. organ
36. system

SYSTEMS OF THE BODY
37. endocrine
38. integumentary
39. skeletal
40. integumentary
41. skeletal
42. lymphatic
43. digestive
44. reproductive
45. lymphatic
46. cardiovascular
47. respiratory
48. muscular
49. nervous
50. muscular
51. nervous
52. endocrine
53. cardiovascular
54. cardiovascular
55. reproductive
56. urinary
57. digestive
58. reproductive
59. urinary

BODY REGIONS (Answers within each region may be in any order. They are placed in alphabetical order here.)
60. cranium
61. face
62. anterior neck
63. posterior neck
64. abdomen
65. back
66. pelvis
67. thorax
68. antebrachium
69. axilla
70. brachium
71. carpus
72. digits (fingers)
73. elbow
74. manus (hand)
75. palm
76. shoulder
77. crus (leg)
78. digits (toes)
79. femoral
80. gluteal
81. knee
82. pes (foot)
83. sole
84. tarsus

BODY CAVITIES
85. false, abdominopelvic
86. true
87. true
88. false, pericardial
89. false, vertebral canal
90. true
91. false, pelvic

DIAGNOSTIC IMAGING
92. computed axial tomography; information from X rays projected from different angles is analyzed by computer to produce images of "slices" of body regions
93. positron emission tomography; a scanner tracks the progress of a radioactive tracer through the body to provide information of physiology

94. echoes of sound waves are collected and analyzed to produced an image
95. magnetic resonance imaging; hydrogen atoms are mapped within the body to produce 3-D images of internal organs

CHARACTERISTICS OF LIFE AND HOMEOSTASIS
96. c & f
97. d & g
98. e & h
99. a & i
100. n
101. o
102. m
103. k
104. j & l
105. b

HEALTH AND DISEASE
106. Disease is any reduction in all body parts functioning cooperatively to maintain homeostasis.
107. A lesion is a structural change in a body part.
108. Acute refers to conditions which are expected to last a relatively short time. Chronic refers to long-lasting conditions.
109. A diagnosis is the identification of disease.
110. A pathologist specializes in diagnoses.
111. Congenital diseases arise before birth. Immunological diseases involve a reaction by the immune system to an invasion.
112. The invasion of the body by a foreign organism.
113. Inflammation is a protective response of the body to invasion often involving redness, heat, swelling, and pain. (This definition came from a dictionary.)
114. A metabolic disease is one which affects metabolism directly.
115. A neoplastic disease is one which involves abnormal growth and reproduction. Tumors and cancer are examples.

INFECTIOUS AGENTS
116. A virus is a noncellular, infectious agent consisting of a nucleic acid core surrounded by a protein coat. Examples are Human Immunodeficiency Virus (HIV) and measles virus.
117. Bacteria are procaryotic (lacking a nucleus) cells. *Clostridium* and *E. coli* are examples.
118. Fungi are a group of simple organisms which obtain their food by absorption. Examples of disease caused by fungi are ringworm and athlete's foot.
119. Protozoa are one-celled organisms which move to obtain food. Examples include the causative agents of malaria and toxoplasmosis.

LABELS AND LISTS
1. A. frontal/coronal
   B. midsagittal
   C. horizontal
2. B
3. A parasagittal plane divides into unequal right and left portion; whereas, a midsagittal plane divides into equal halves.
4. chemical, cell, tissue, organ, system, organism
5. proteins, fats, carbohydrates, nucleic acids
6. epithelial, muscular, connective, nervous
7. integumentary, muscular, skeletal, nervous, endocrine, cardiovascular, lymphatic, respiratory, digestive, urinary, reproductive
8. See your textbook for the correct answers.
9. See your textbook for the correct answers.
10. See your textbook for the correct answers.
11. organization, metabolism, movement, excitability, growth, reproduction
12. congenital, immunological, metabolic, neoplastic

LABELS AND LISTS
1. A. frontal/coronal
   B. midsagittal
   C. horizontal
2. B
3. A parasagittal plane divides into unequal right and left portion; whereas, a midsagittal plane divides into equal halves.
4. chemical, cell, tissue, organ, system, organism
5. proteins, fats, carbohydrates, nucleic acids
6. epithelial, muscular, connective, nervous
7. integumentary, muscular, skeletal, nervous, endocrine, cardiovascular, lymphatic, respiratory, digestive, urinary, reproductive
8. See your textbook for the correct answers.
9. See your textbook for the correct answers.
10. See your textbook for the correct answers.
11. organization, metabolism, movement, excitability, growth, reproduction
12. congenital, immunological, metabolic, neoplastic

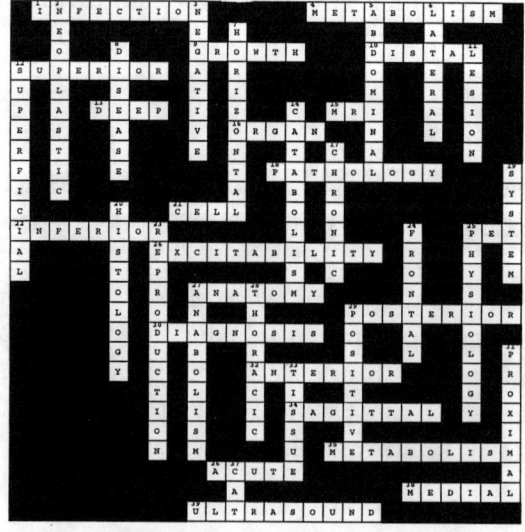

ANSWERS - CHAPTER 2

COMPOSITION OF MATTER
1. matter
2. mass
3. atoms
4. element
5. 109
6. chemical symbol
7. nucleus, electron
8. neutrons, protons
9. electrons
10. electron
11. electron
12. electron orbital
13. +
14. -
15. neutron
16. neutral
17. ion
18. atomic number
19. atomic weight
20. molecule
21. atom
22. molecule
23. K
24. K^+
25. Ca^{2+}
26. Na^+
27. one
28. two
29. cation
30. gained
31. three
32. four

CHEMICAL BONDS
33. molecule
34. chemical bond
35. chemical bond
36. electron shell
37. inert
38. reactive
39. ion
40. ionic bond
41. cations
42. anions
43. covalent bond
44. single covalent bond
45. double covalent bond
46. hydrogen bonds

CLINICAL USES
47. isotope
48. radioisotopes
49. radioactive decay
50. cancer
51. Co (cobalt)
52. may
53. can
54. more
55. synthesis produces larger molecules; decomposition results in simpler molecules
56. Glycerol and fatty acids are the reactants; fat is the product.
57. repair
58. end products are CO_2 & H_2O
59. anabolism

CHEMICAL COMPOUNDS...
60. false, organic
61. true
62. false, water
63. true
64. false, water
65. true
66. false, small
67. false, slowly
68. true
69. false, electrolytes
70. false, acid
71. true
72. true
73. buffer systems
74. true
75. false, neutral
76. true
77. true
78. pH 2.7
79. pH 11
80. acid
81. base
82. water
83. carbon
84. ions
85. buffer
86. water
87. water
88. electrolyte
89. heat capacity

ORGANIC COMPOUNDS
90. organic
91. DNA
92. carbohydrates
93. fats
94. proteins
95. phospholipid
96. lipids, fats, and steroids
97. RNA
98. fats
99. nucleotides
100. nucleotides
101. protein
102. phospholipid
103. steroid
104. polysaccharide
105. lipids
106. DNA
107. ATP
108. protein
109. DNA & RNA

COMPARE NUCLEIC ACIDS
110. deoxyribose
111. two
112. A, T, G, C
113. ribose
114. one
115. A, U, G, C

MORE REVIEW!
116. amino acids
117. nucleotides
118. atoms
119. protons, neutrons, and electrons
120. molecules
121. amino acids

LABELS AND LISTS
1. See your textbook for the correct answers.
2. C, H, O, N
3. C, H, O, N, Ca, P
4. covalent, ionic, hydrogen
5. universal solvent, transport medium, high heat capacity, effective lubricant
6. proteins, lipids, carbohydrates, nucleic acids
7. carbohydrates: C, H, O
   lipids: C, H, O
   proteins: C, H, O, N, S
   nucleic acids: C, H, O, N, P

REVIEW, REVIEW
1. The planetary model depicts electrons revolving around in the nucleus in a circular orbit. The orbital model depicts the region around the nucleus where an electron is most likely to be found.
2. The orbital model is more accurate.
3. The planetary model is used for the sake of simplicity.
4. Atoms have an equal number of electrons and protons so the positive charges are balanced our by the negative charges.
5. An ion is an atom with a charge, due to the fact that it has gained or lost electrons(s).
6. "Molecules have their own set of properties, which are quite different from the properties of the atoms which make them up."
7. The term "energy level" and "electron shell" are synonymous. They can be used interchangeably.
8. Covalent bonds involve the sharing of electrons by atoms; whereas, ionic bonds involve the loss/ gain of electrons.
9. Milk of magnesia is a base (pH 10.0 - 11.0) which neutralizes the stomach acid.
10. ATP is more similar to RNA than to DNA, because it is made from ribose sugar.

ANSWERS - CHAPTER 3

CELLULAR ENVIRONMENT
1. intracellular
2. extracellular
3. extracellular fluid
4. secretion
5. interstitial fluid
6. plasma
7. interstitial fluid
8. plasma
9. interstitial fluid
10. matrix
11. support
12. intracellular fluid
13. plasma membrane
14. intracellular environment
15. intracellular fluid
16. cytoplasm
17. organelles
18. nucleus

WORD STUDY
19. intracellular fluid (ICF)
20. extracellular fluid (ECF)
21. interstitial fluid
22. protoplasm

REVIEW
23. intracellular environment
24. intercellular environment
25. interstitial
26. intracellular
27. protoplasm
28. plasma membrane
29. cytoplasm

CELLULAR STRUCTURE
30a. barrier for cell and regulates movement of materials in and out of the cell
30b. around cell
30c. region of cellular metabolism
30d. between plasma membrane and nucleus
30e. pathway for movement within the cell, RER is site of protein synthesis, SER is site of lipid synthesis
30f. throughout cell often connecting nucleus and plasma membrane
30g. prepares and packages secretions
30h. usually near nucleus
30i. primary site of ATP production
30j. throughout cytoplasm
30k. "digestive" enzymes, suicide sac
30l. throughout cytoplasm
30m. detoxify certain toxins
30n. throughout cytoplasm
30o. structural support and movement
30p. skeleton between organelles, extensions into cilia and flagella for movement
30q. contains DNA which functions to control activities of the cell
30r. often near center of cell
30s. regulates movement between nucleus and cytoplasm
30t. surrounds nucleus
30u. site of DNA
30v. within nucleus
30w. site of RNA synthesis
30x. within nucleoplasm
30y. contains DNA (genes)
30z. throughout nucleoplasm

PLASMA MEMBRANE
31. d
32. a
33. b
34. a
35. c
36. d
37. b
38. b
39. a, b, c, d
40. c
41. a
42. a
43. a
44. d
45. a
46. a
47. a
48. b
49. c
50. b
51. 1 - lumen
    2 - free surface
    3 - lateral border
    4 - basal surface
    5 - basement membrane
52. ATP
53. against
54. cytosis
55. endocytosis
56. exocytosis
57. pseudopod
58. phagocytosis
59. pinocytosis
60. exocytosis
61. excretion
62. secretion

SHORT ANSWER

63. Passive processes use kinetic energy to move substances across the membrane, thus the cell need not expend any ATP. Active processes require the use of ATP.
64. Diffusion, facilitated diffusion, and osmosis all occur with the concentration gradient and therefore require no ATP. All active transport processes such as endocytosis and exocytosis work against concentration gradients and so require ATP.
65. Both diffusion and facilitated diffusion are the movement of chemicals from an area of high concentration to an area of low concentration. The difference is in the fact that facilitated diffusion involves a membrane protein which provides a channel or pathway for diffusion to occur. Regular diffusion occurs through the lipid bilayer without the use of a protein channel.
66. Osmosis is the diffusion of water across a semipermeable membrane.
67. Distilled water is hypotonic (lower concentration of solutes) compared to the cells. Water will move into the cells along the concentration gradient of water by osmosis. Eventually, this could lead to cell death, but homeostasis is maintained by the reabsorption of salts and the excretion of excess water from the kidneys.
68. The cell (see #67).
69. The cells would expand and perhaps undergo hemolysis.
70. Isotonic solutions have the same osmotic pressure as the body fluids and so result in less cell damage.
71. The bacterium would be digested by enzymes.
72. Both processes are active, requiring ATP.
73. Exocytosis removes material from the cell; endocytosis takes material into the cell.

74. cytoplasm
75. cytosol
76. organelles
77. endoplasmic reticulum
78. rough endoplasmic reticulum
79. smooth endoplasmic reticulum
80. ribosomes
81. RER
82. liver
83. Golgi apparatus
84. mitochondria
85. cellular respiration
86. mitochondrion
87. cristae
88. surface area
89. catabolic enzymes
90. DNA

91. protein
92. muscle cells
93. lysosomes
94. digestive system of the cell
95. self destruct bags
96. suicide packets
97. enzymes
98. Tay-Sachs
99. peroxisomes
100. cytoskeleton
101. is not
102. microtubules
103. microfilaments
104. myofilaments
105. nucleus
106. DNA (the genetic material)
107. kernel
108. nuclear membrane

109. nucleoplasm
110. pores
111. nucleoli
112. chromatin
113. ribosome synthesis
114. chromosomes
115. genes

MATCHING

116. plasma membrane
117. lysosomes
118. ER
119. Golgi apparatus
120. cytoplasm
121. cytosol
122. organelles
123. mitochondria
124. peroxisomes
125. cytoskeleton
126. nucleus

127. nuclear membrane
128. nucleoplasm
129. nucleoli
130. chromatin

MORE REVIEW
131. Golgi apparatus
132. Golgi apparatus
133. rough endoplasmic reticulum (enzymes are proteins) and Golgi apparatus (secretions)
134. smooth endoplasmic reticulum (steroid are lipids)
135. mitochondria (to meet energy requirement)

PROTEIN SYNTHESIS
136. DNA
137. transcription and translation
138. genetic code
139. three
140. triplet code
141. gene
142. mRNA
143. transcription
144. through pores
145. codon
146. translation
147. anticodon
148. ribosomes
149. two
150. stop

151. Different genes located on DNA are transcribed at different times. Transcription is necessary before proteins can be manufactured.
152. DNA carries of series of triplet codes, each composed of three nucleotide base pairs, which each code for a specific amino acid.
153. A gene codes for one polypeptide chain or protein. Each polypeptide chain is coded for by one gene.
154. Messenger RNA (mRNA) is manufactured in the nucleus during the process of transcription. The triplet codes are transcribed into codons of mRNA. The mRNA carries the codons to the ribosomes for protein translation. The ribosome is composed in part of rRNA and functions as the protein factory. tRNA functions in the cytoplasm to pick up specific amino acids which it brings to the ribosomes. The codon-anticodon base pair matching assures that he appropriate amino acid is put in place.
155. The triplet code nucleotides are the complements of the nucleotides of the codons. For example, if a triplet code is GCG, then the codon is CGC. If the triplet code is AAA, the codon is UUU. (Uracil in RNA instead of the thymine if it were DNA base pair matching.)
156. Transcription involves "writing" the genetic code in RNA nucleotides instead of DNA nucleotides. This occurs in the nucleus. The result of transcription is mRNA. Translation puts the code in terms of amino acids, it changes "languages". The result is an amino acid chain.

157. d
158. d
159. b
160. a
161. b
162. c
163. d
164. e
165. a
166. d
167. b
168. b
169. a
170. a

LABELS AND LISTS
1. plasma membrane, cytoplasm, nucleus
2. microvilli, cilia, flagellum. diffusion, facilitated diffusion, osmosis, filtration
3. diffusion, facilitated diffusion, osmosis, filtration
4. ER, Golgi apparatus, lysosomes, peroxisomes, mitochondria
5. prophase, metaphase, anaphase, telophase (PMAT)
6. - 10. See your textbook.

ANSWERS - CHAPTER 4

TYPE	STRUCTURE	EXAMPLE
Simple squamous	single layer, flat cells	lining blood and lymph vessels, air sacs of lungs, and body cavities
Simple cuboidal	single layer, cube-shaped cells	walls of ducts in kidneys, liver and many glands
Simple columnar	single layer, elongated cells	line uterus, stomach, and small intestine
Stratified squamous	more than one layer, surface cells squamous (flat), deeper cells may be cuboidal or columnar	epidermis (outermost layer of skin)
Pseudostratified columnar	appears to be many layered, but is single layer; cells of various sizes, but all touch basement membrane	lining bronchi and trachea
Transitional	many-layers of cube-like or irregular cells	lining urinary bladder

FILL IN THE BLANKS
1. avascular
2. glandular
3. basement membrane
4. simple squamous
5. stratified squamous
6. pseudostratified
7. transitional
8. glandular
9. exocrine
10. endocrine

CONNECTIVE TISSUE
11. collagenous fibers
12. collagen
13. collagen
14. collagen disease
15. scurvy
16. elastic fibers
17. elastin
18. reticular fibers
19. fibroblast
20. loose connective tissue
21. loose connective tissue
22. adipose tissue
23. adipocytes
24. dense connective tissue
25. dense irregular connective tissue
26. dense regular connective tissue

MATCHING
27. cartilage
28. matrix
29. chondrocyte
30. lacunae
31. perichondrium
32. hyaline cartilage
33. elastic cartilage
34. fibrocartilage
35. elastic cartilage
36. hyaline cartilage
37. fibrocartilage
38. bone
39. osteocytes
40. lacunae
41. periosteum
42. compact bone
43. osteonic canals
44. lamellae
45. canaliculi
46. osteon
47. spongy bone
48. red marrow
49. spicules

BLOOD-FORMING
TISSUE AND BLOOD
50. true
51. true
52. true
53. false, in cavities of spongy bone
54. true
55. false, lymphoid tissue
56. false, connective
57. true
58. false, platelets
59. true

MUSCLE AND NERVE TISSUE
60. d
61. c
62. a
63. a
64. e
65. b
66. c
67. b
68. a
69. d
70. a
71. b

TISSUES, TUMORS, AND CANCER
72. A tumor is an overgrowth of cells which serve no useful purpose.
73. Neoplasm and tumor mean the same thing.
74. A carcinogen is something in the environment which causes cancer-producing mutations. Two examples are x-rays and the chemicals in cigarettes.
75. Benign tumors grow slowly and remain in one location. Malignant tumors grow rapidly and have the ability to move to new locations.
76. Metastasis is the spreading of a cancer to another location.
77. Malignant tumors are commonly known as cancers.
78. See Table 4-3.
79. Do not use tobacco. Avoid prolonged, unprotected exposure to the sun. Avoid unnecessary X-rays. Avoid fatty, smoked, or cured food. Increase her fiber intake. Examine herself for suspicious mole or lumps.

MEMBRANES
80. cutaneous
81. synovial
82. mucous
83. serous

LABEL AND LIST
1. epithelial, connective, muscular, nervous
2. any two: simple squamous, simple cuboidal, simple columnar, stratified columnar, pseudostratified columnar, transitional
3. any two: loose (areolar), adipose, dense (regular or irregular), cartilage (hyaline, elastic, or fibrocartilage), bone (spongy or compact), hemopoietic tissue (blood-forming), blood

4.

Type of Muscle Tissue	Type of Nervous Control	Microscopic Appearance	Example
Skeletal	Voluntary	Striated	Muscles of arm
Smooth (visceral)	Involuntary	Non-striated	Muscle of stomach
Cardiac (myocardium)	Involuntary	Striated with intercalated disks	Muscle of heart

5.

Membrane	Location
A. Cutaneous	Skin
B. Serous	Lining internal surfaces of thoracic and abdominal cavities.
C. Mucous	Lining internal surfaces of digestive, respiratory, reproductive, and urinary tracts.
D. Synovial	Lining internal surfaces of freely moveable joints.

6. carcinoma, sarcoma, leukemia

Crossword Solution

Across:
- 3. SKELETAL
- 5. FIBROCARTILAGE
- 7. OSTEON
- 8. NERVE
- 9. COMPACT
- 10. STRATIFIED
- 12. CHONDROCYTE
- 14. ELASTIC
- 18. OSTEOCYTE
- 20. PLASMA
- 21. ELASTIN
- 24. PSEUDOSTRATIFIED
- 26. DENSE
- 27. AREOLAR
- 28. CANALICULI
- 29. COLUMNAR
- 32. MUSCLE
- 33. CONNECTIVE
- 34. SEROUS
- 35. TISSUE
- 36. LACUNA
- 37. SYNOVIAL
- 38. ENDOCRINE

Down:
- 1. SIMPLE
- 2. TRANSITION
- 3. SMOOTH
- 4. AXON
- 6. ADIPOSE
- 7. OSTEOBLAST (OSTEO...)
- 10. SPONGING / SPONGY
- 11. LAMELLAE
- 13. HEMATOMA
- 15. CILIA
- 16. EPITHELIAL
- 17. MUCOUS
- 18. OSTEOCLAST
- 19. SUBMUCOSA
- 22. SQUAMOUS
- 23. OSTEOTONIC
- 25. RETICULAR
- 26. DETAILED
- 27. ADIPOSE
- 30. CARTILAGE
- 31. HYALINE
- 32. MEMBRANE
- 33. COLLAGENOUS
- 34. SETTLE

ANSWERS - CHAPTER 5

INTEGUMENTARY SYSTEM
1. skin (or integument or cutaneous membrane)
2. integument, cutaneous membrane
3. epidermis
4. dermis
5. hypodermis
6. epidermis
7. glands
8. glands
9. blood vessels
10. receptors
11. homeostasis

SKELETAL SYSTEM
12. bones
13. true
14. skeleton
15. 206
16. axial skeleton
17. appendicular
18. appendicular
19. axial
20. appendicular
21. axial
22. joint
23. protection

MUSCULAR SYSTEM
24. true
25. false, skeletal only
26. false, 500
27. false, bundles
28. true
29. false, nerve cells
30. false, skeletal movement

NERVOUS SYSTEM
31. peripheral nervous system
32. efferent
33. toward
34. away from
35. somatic nervous system
36. autonomic nervous system
37. neurons
38. homeostasis

ENDOCRINE SYSTEM
39. ducts
40. pituitary gland
41. thyroid gland
42. parathyroid glands
43. adrenal glands
44. pancreas
45. gonads
46. thymus gland
47. pineal gland
48. stomach and kidney
49. hormones
50. diffusion
51. target
52. homeostasis

CARDIOVASCULAR SYSTEM
53. arteries
54. veins
55. capillaries
56. cardiac muscle
57. atria
58. ventricles
59. atria
60. circulatory system
61. formed elements
62. plasma
63. erythrocytes
64. leukocytes
65. thrombocytes
66. erythrocytes
67. erythrocytes
68. thrombocytes

LYMPHATIC SYSTEM
69. system of tubes which transports fluid
70. lymphatic system has no pump and only moves fluid toward the heart
71. lymph
72. monocytes and lymphocytes
73. to protect the body from harmful microorganisms
74. a. stationary cells filter out harmful agents as the lymph passes toward the heart; b. travelling cells actively pursue harmful materials and destroy them
75. a systemic disease is one which affects the entire body
76. systemic lupus erythematosus (SLE), tuberculosis (TB), cystic fibrosis (CF), acquired immunodeficiency syndrome (AIDS), many forms of cancer, etc.

RESPIRATORY SYSTEM
77. false, carbon dioxide
78. true
79. false, pharynx
80. False, larynx
81. true
82. false, bronchi
83. true
84. true

DIGESTIVE SYSTEM
85. mouth
86. salivary glands
87. esophagus
88. stomach
89. small intestine
90. bile
91. gallbladder
92. liver
93. pancreas
94. small intestine
95. large intestine
96. anus

URINARY SYSTEM
97. kidneys
98. ureters
99. urinary bladder
100. urethra
101. salt
102. urine

REPRODUCTIVE SYSTEM
103. testes
104. sperm
105. sperm
106. scrotum
107. vas deferens
108. urethra
109. semen
110. prostate gland
111. ovaries
112. ovum
113. fallopian tube
114. uterus
115. vagina
116. vulva

LABELS AND LISTS
1. (in alphabetical order)
cardiovascular
digestive
endocrine
integumentary
lymphatic
muscular
nervous
reproductive
respiratory
skeletal
urinary

2. support, protection, storage of mineral salts, site of blood cell formation, attachment site for muscles (via tendons)
3. skeletal movement, maintenance of posture, support of the skeleton, produces heat
4. central nervous system (CNS) and peripheral nervous system (PNS)
5. lymphatic vessels, lymph nodes, spleen, thymus gland, tonsils

ANSWERS - CHAPTER SIX

FUNCTIONS OF THE INTEGUMENTARY SYSTEM
1. e
2. d
3. c
4. d
5. c
6. a
7. b
8. d
9. c
10. c

THE EPIDERMIS
11. stratum basale
12. stratum spinosum
13. stratum lucidum
14. stratum corneum
15. melanocyte
16. keratin
17. stratum basale
18. stratum corneum
19. stratum spinosum
20. stratum corneum
21. stratum lucidum
22. four, except in soles and palms - five
23. four
24. melanocytes
25. tanning
26. malignant melanoma
27. melanin
28. carotene
29. blood vessels
30. squamous cell carcinoma
31. basal cell carcinoma
32. malignant melanoma

THE DERMIS
33. connective
34. collagen
35. 25
36. temperature
37. papillary; reticular
38. papillae
39. friction ridges (fingerprints and footprints)
40. finger
41. little network
42. thicker
43. strength; elasticity
44. wrinkles

ACCESSORY ORGANS

45. epidermal derivatives arise from epidermal cells during embryonic development, though they may end up in the dermis of the adult
46. cell division, the production of new cells, occurs at the base of the hair, pushing older cells outward
47. The cells of the hair shaft are dead cells.
48. The arrector pili muscles pull the hairs into an upright position. This produces a small amount of heat to warm the body and traps some air under the hair shafts, providing insulation.
49. Sebum is an oily substance produced by sebaceous glands. It helps waterproof the skin, and keeps hairs and the skin pliable.
50. A plugged sebaceous gland duct is known as a blackhead.
51. Eccrine glands are located everywhere across the surface of the skin and produce a watery sweat to cool the body. Apocrine glands are located primarily in the armpits and groin regions, and produce a sweat with a higher protein content.
52. false, 1 mm/3 days
53. false, no effect
54. false, 100
55. true, promotes hair loss
56. false, stimulate
57. false, melanin
58. true
59. true
60. true, sebaceous glands are associated with hairs; there are no hairs on the palms or soles
61. true
62. true
63. false, sudoriferous
64. true
65. false, apocrine glands
66. false, eccrine

NAILS
67. nail body
68. eponychium
69. nail root
70. lunula
71. brittle nails

RECEPTORS
72. distal
73. toward the brain
74. sensations
75. Pacinian corpuscles
76. Meissner's corpuscles

SKIN REPAIR
77. b
78. d
79. a
80. b

HYPODERMIS
81. false, deep
82. true
83. true
84. true
85. true

86. at lower temperatures, enzyme activity slows or stops; at higher temperatures, enzymes can denature (breakdown)

87. c. receptors in skin sense the external heat and send message to brain
    b. brain stimulates sweat glands to secrete
    d. brain signals blood vessels in dermis to increase blood flow to the skin
    f. excessive heat in blood passes through the skin to the air
    a. sweat evaporates and reduces the temperature in skin and nearby blood
    e. cooled blood circulates through body to lower internal temperature

88. Gertrude's blood vessels have dilated (expanded) to allow more blood to circulate closer to the surface where it can radiate excess heat and be cooled by the evaporating sweat.

89. involuntary contractions of the skeletal muscle

90. contractions of muscles, skeletal and arrector pili, produce heat as a by product of contraction; "goosebumps" are the result of the hair shafts pushing the skin into "bumps" when the arrector pili muscle pulls on them

91. hot blood from the deeper areas of the body is moved more rapidly to the surface where it can cool

92. constrict, to reduce the amount of blood, and thereby heat, that is exposed to the cold weather

CLINICAL TERMS AND DISEASES
93. acne vulgaris
94. Herpes simplex; Herpes zoster
95. dermatitis
96. boil
97. ringworm
98. wart
99. psoriasis
100. tumor
101. Herpes zoster
102. pediculosis
103. boil, carbuncle
104. carbuncle
105. burn

DEAR DOCTOR
106. b
107. j
108. k
109. i
110. h
111. c
112. c
113. g
114. e
115. f
116. a
117. a
118. d

LABELS AND LISTS
1. protection, regulation of temperature, sensory perception, excretion of wastes, production of vitamin D
2. epidermis, dermis
3. strata basale, spinosum, granulosum, lucidum, corneum
4. squamous cell carcinoma, basal cell carcinoma, malignant melanoma

ANSWERS - CHAPTER 7

BONE STRUCTURE
1. long bone
2. flat bone
3. irregular bone
4. short bones
5. diaphysis
6. articular cartilage
7. joint
8. epiphyses
9. periosteum
10. compact bone
11. spongy bone
12. medullary cavity
13. osteoblast
14. osteocytes
15. osteoclasts
16. osteoblast
17. osteocyte
18. osteon
19. Volkmann's canal
20. red marrow

BONE DEVELOPMENT...
21. embryonic membrane, hyaline cartilage, lengthwise, width wise, recycle bone matrix
22. about fifth week of life
23. within a membrane
24. within cartilage
25. osteoblasts secrete new matrix
26. secretion of new bone matrix
27. produce matrix
28. a. osteoblasts
    b. osteoblasts
    c. fifth week
    d. sixth week
    e. osteoblast secretes matrix along embryonic membranes
    f. chondroblast produces cartilage model which is replaced by bony matrix
29. chondrocytes
30. chondrocytes
31. diaphysis
32. epiphysis
33. epiphyseal plate
34. articular cartilage
35. interstitial growth
36. epiphyseal plate
37. epiphyseal line
38. appositional growth
39. osteoclasts
40. true
41. false, bone continues to remodel
42. false, reabsorption is done by osteoclasts
43. false, remodeling occurs all of the time, though not in every part of every bone
44. true
45. true
46. false, the break extends partially through the bone
47. false, these are normal
48. false, it is a procallus
49. true
50. false, another name is "simple"

ORGANIZATION...
51. appendicular, axial
52. 206
53. true
54. no answers! This is something to do.
55. 22
56. sutures
57. sinuses
58. sinusitis
59. frontal
60. orbit
61. parietals
62. foramen magnum
63. atlas
64. external auditory meatus
65. zygomatic
66. hyoid
67. sella turcica
68. sella turcica
69. maxillary sinuses
70. alveolar process
71. cleft palate
72. vomer
73. turbinates
74. mandible
75. hyoid
76. 7
77. 12
78. 5
79. 5
80. 3 - 5
81. false, larger
82. false, thoracic
83. true
84. false, strength
85. true
86. false, kyphosis is hunchback, lordosis is swayback
87. true
88. true
89. false, men have 24 also
90. false, do not directly articulate with the sternum
91. false, only 14!, (thumb) has 2
92. true
93. true

ARTICULATIONS
94. fibrous, fibrous material, little to no movement, suture
95. cartilaginous, bound with cartilage, slightly moveable, discs
96. synovial, fluid-filled space between bones, freely moveable, elbow
97. fontanels
98. syndesmosis
99. gomphosis
100. cartilaginous
101. cartilaginous
102. herniated disc
103. laminectomy
104. synovial joint
105. synovial cavity
106. cartilaginous
107. suture
108. bursa
109. tendon sheath
110. gliding
111. ball and socket
112. saddle
113. pivot
114. saddle
115. condyloid
116. gliding
117. pivot
118. ball and socket

MOVEMENTS
119. flexion
120. dorsiflexion
121. extension
122. hyperextension
123. plantar flexion
124. abduction
125. adduction
126. circumduction
127. rotation
128. pronation
129. supination
130. eversion
131. inversion
132. protraction
133. retraction

BASEBALL
134. rotation
135. hyperextension
136. flexion
137. extension
138. abduction
139. adduction
140. protraction
141. circumduction

CLINICAL TERMS
142. osteomalacia, rickets
143. osteoarthritis
144. gout
145. achondroplasia
146. osteitis deformans
147. rickets
148. rheumatoid arthritis
149. tumors of bone
150. osteoporosis
151. osteomyelitis
152. osteomalacia
153. osteoporosis
154. osteomyelitis

LABELS AND LISTS
1. support, protection, aid in movement, blood cell formation, storage of minerals
2. long, short, flat, irregular
3. osteoblasts, osteocytes, osteoclasts
4. intramembranous, endochondrial
5. frontal, mandible, parietal, clavicle, occipital, temporal
6. femur, tibia, fibula (or any three appendicular bones)
7. immovable, slightly moveable, freely moveable
8. calcium phosphate and calcium carbonate
9. See your textbook for the correct answers.
10. See your textbook for the correct answers.
11. See your textbook for the correct answers.
12. skull, vertebral column, thoracic cage, hyoid
13. pectoral girdles, upper limbs, pelvic girdles, lower limbs
14. fossa, foramen, fissure, condyle, tubercle, trochanter, tuberosity, facet, process, spine. See Table 7-1.
15. drain fluids, reduce weight of skull, resonate sound of voice
16. frontal, ethmoid, sphenoid, two maxillary bone
17. see Table 7-2
18. ilium, pubis, ischium
19. humerus, ulna, radius, carpals, metacarpals, phalanges
20. fibrous, cartilaginous, synovial
21. gliding, hinge, pivot, condyloid, saddle, ball and socket
22. A - abduction, B - pronation, C - inversion, D - flexion, E - flexion, F - supination, G - hyperextension, H - adduction
23. support, protection, blood cell formation, mineral recycling

TEST YOURSELF
1. e
2. a
3. b
4. e
5. a
6. d
7. c
8. a
9. a
10. b
11. d
12. b
13. e
14. b
15. a
16. e
17. e
18. c
19. d
20. a
21. e
22. c
23. d
24. b
25. a
26. d
27. e
28. a
29. e
30. a

Completed crossword puzzle.

ANSWERS - CHAPTER 8

MUSCLE STRUCTURE
1. contractility
2. excitability
3. extensibility
4. elasticity
5. skeletal
6. deep fascia
7. deep fascia
8. epimysium
9. tendon
10. aponeurosis
11. surgery
12. tendinitis
13. fiber
14. sarcolemma
15. sarcoplasm
16. mitochondria
17. sarcoplasmic
18. transverse
19. myofibrils
20. thin filaments
21. myosin
22. actin
23. A
24. sarcomere
25. resting membrane potential
26. neuromuscular junction
27. synaptic vesicles
28. action potential
29. motor unit
30. neurotransmitter
31. acetylcholine

PHYSIOLOGY OF MUSCLE CONTRACTION
32. d
33. b
34. d
35. b, d, c, a
36. c, b, a, d
37. c
38. a
39. d
40. c
41. c
42. a
43. a
44. a
45. b
46. a

	SKELETAL	SMOOTH	CARDIAC
47. Where is it found?	attached to skeleton	in walls of hollow organs (e.g. digestive system, blood vessels)	walls of heart
48. How is it controlled	voluntary	involuntary	involuntary
49. Describe the shape of the fibers	long, cylindrical	spindle-shaped	cylindrical, branching
50. Are there striations?	yes	no	yes
51. How would they place in a contraction speed race?	first place	third place	second place
52. How would they place in a strength competition?	first place	third place	second place
53. Which can stay contracted the longest time?		THE WINNER!	

MUSCLE RESPONSES
54. threshold stimulus
55. subthreshold stimulus
56. all-or-none response
57. recruitment
58. twitch
59. treppe
60. wave summation
61. complete tetanus
62. muscle tone
63. isotonic
64. isometric
65. motor skill development
66. endurance

PRODUCTION OF MOVEMENT
67. b
68. d
69. a
70. c
71. f
72. g
73. e

NAMING THE MAJOR MUSCLES
74. Orbicularis oculi
75. Orbicularis oris
76. Buccinator
77. Zygomaticus
78. Masseter
79. Sternocleidomastoid
80. Trapezius
81. Levator scapulae
82. Serratus
83. Pectoralis minor
84. Deltoid
85. Subscapularis
86. Supraspinatus
87. Teres
88. Triceps brachii
89. Biceps brachii
90. Rectus abdominis

91. Iliopsoas
92. Gluteus maximus
93. Gracilis
94. Tensor facia latae
95. Sartorius
96. Quadriceps femoris
97. Gastrocnemius
98. Soleus
99. Peroneus longus

INTERESTING INFO
100. rest, ice, compression, elevation
101. orbicularis oris, buccinator
102. trapezius
103. latissimus dorsi
104. latissimus dorsi
105. deltoid
106. true
107. linea alba
108. gluteus medius
109. hamstring
110. Achilles' tendon

HOMEOSTASIS
111. Movement, heat
112. Manipulation of the environment for eating etc., as well as assisting in blood flow to the heart
113. 80% of body heat is derived from energy released from ATP during muscle contractions, this keeps the body warm

CLINICAL TERMS
114. b
115. a
116. c
117. muscular dystrophy
118. myoma
119. myositis
120. myotonia
121. shinsplints
122. torticollis

LABELS AND LISTS
1. contractility, excitability, extensibility, elasticity
2. movement, support, production of heat
3. Ca ions free binding sites on actin, ATP is cleaved to release energy, crossbridges form between actin and myosin. There are other requirements for contraction as well, such as: neurotransmitter release at neuromuscular junction, action potential transmission along muscle fiber surfaces and transverse tubules, and Ca ion release.
4. - 14. See the figures and tables in your textbook.

QUESTIONS TO MAKE YOU THINK
1. Deep fascia is composed of dense irregular connective tissue. This tissue contains a large amount of collagen fibers which are flexible and yet have great tensile strength.
2. Neurotransmitters cross the synaptic cleft by diffusion.
3. An ion is an atom which has gained (anion) or lost (cation) an electron or electrons. Calcium ions are cations and have a positive charge. Calcium ions are actively transported across the membrane of the sarcoplasmic reticulum.
4. The all-or-none principle applies to motor units - a nerve cell and the muscle cells it innervates. A muscle is made up of many motor units. Not all motor units contract at the same time; though, the muscle cells within a motor unit contract completely or ot at all.

5. Normal contractions of skeletal muscles are dependent upon nerve impulses and the neurotransmitters that are released into the synaptic clefts. If the nerve is damaged, the muscle does not receive the signal to contract and it remains relaxed; the eye remains open.

ANSWERS - CHAPTER 9

DIVISIONS OF THE NERVOUS SYSTEM
1. central nervous system
2. brain
3. sensory
4. originate in
5. peripheral nervous system
6. cranial
7. spinal
8. somatic
9. autonomic
10. involuntary
11. sympathetic
12. sympathetic
13. parasympathetic

NERVE TISSUE
14. neurons
15. neuroglia
16. neurons
17. neuroglia
18. neuroglia
19. astrocytes, ependymal cells, microglia, & oligodendrocytes
20. Schwann cells
21. astrocytes
22. ependymal cells
23. microglia
24. oligodendrocytes
25. Schwann cells
26. cell body
27. nissl bodies
28. neurofibrils
29. dendrites
30. axon
31. dendrites
32. axon
33. myelin sheath
34. dendrite
35. neurilemma
36. nodes of Ranvier
37. nodes of Ranvier
38. unmyelinated fibers
39. white matter
40. gray matter

TYPES OF NEURONS
41. multipolar
42. bipolar
43. unipolar
44. afferent neuron
45. CNS
46. efferent neuron
47. motor neurons

NEURON FUNCTION
48. b
49. d
50. b
51. b
52. a
53. a
54. positive (inside negative!)
55. d
56. d
57. a
58. a
59. b
60. d
61. b
62. d
63. b
64. c
65. d
66. c
67. e, a, d, b, f, c
68. a
69. a
70. c
71. d
72. c, d, e
73. a

CHEMICALS AND THEIR INFLUENCES
74. c
75. d
76. e
77. b
78. a

THE SPINAL CORD
79. meninges
80. dura mater
81. epidural space
82. subarachnoid space
83. pia mater
84. subarachnoid space
85. cerebrospinal fluid
86. pia mater
87. dura mater
88. spinal tap
89. the conus medullaris is cone-shaped and located in the middle of the back
90. the cauda equina is formed of many spinal nerves travelling down from the conus medullaris; it looks like a horse's tail
91. cerebrospinal fluid

92.

SPINAL CORD HORNS	FUNCTION
Anterior gray horn	contains cell bodies of motor neurons
Posterior gray horns	terminal endings of sensory neurons
Lateral gray horns	cell bodies of some motor neurons of the autonomic nervous system

93. myelinated
94. ascending
95. descending
96. reflex arc
97. sensory neuron
98. association neuron
99. somatic reflex
100. somatic reflex
101. visceral reflex

THE BRAIN
102. cerebrum
103. diencephalon
104. pons
105. medulla oblongata
106. cerebellum
107. C
108. G
109. H
110. E
111. F
112. B
113. D
114. A
115. I
116. a
117. b (by the choroid plexus)
118. d
119. a
120. a
121. c, b, e, h, g, f, a, d
122. c
123. a
124. cerebrum

125. cerebellum
126. diencephalon
127. medulla oblongata
128. pons
129. midbrain
130. higher brain
131. cerebral hemispheres
132. convolutions
133. gyri
134. sulci
135. longitudinal fissure
136. transverse fissure
137. lobes
138. gray
139. corpus callosum
140. diencephalon
141. thalamus, hypothalamus
142. gray
143. double brain
144. thalamus
145. pineal gland
146. hypothalamus
147. thalamus
148. thalamus
149. hypothalamus
150. hypothalamus
151. hypothalamus
152. hypothalamus
153. hypothalamus
154. hypothalamus
155. midbrain
156. cerebral peduncles
157. colliculi
158. pons

159. pons
160. cerebellum
161. medulla oblongata
162. folia
163. vermis
164. arbor vitae
165. sulci
166. cerebellar peduncles
167. cerebellum
168. brain stem
169. medulla oblongata
170. medulla oblongata

THE PERIPHERAL
NERVOUS SYSTEM
171. somatic system
172. autonomic system
173. somatic
174. autonomic
175. mixed nerves
176. nerve fiber
177. nerve
178. epineurium
179. epineurium
180. endoneurium
181. perineurium
182. sensory nerve
183. sensory nerve
184. motor nerve
185. motor nerve
186. motor nerve
187. mixed nerve
188. ganglion (singular of ganglia)

CRANIAL NERVES
189. 12 pairs
190. The Roman numerals correspond to the order in which the cranial nerves arise from the brain. The Olfactory Nerve I is the most superior and anterior; the Hypoglossal Nerve XII is most inferior.
191. Olfactory I, Optic II, and Vestibulocochlear VIII
192. Accessory XI and Hypoglossal XII

SPINAL NERVES
193. thirty-one
194. thousands
195. eight
196. twelve
197. five
198. five
199. one
200. anterior horns
201. posterior horns
202. dorsal roots
203. ventral roots
204. dorsal root ganglion
205. dorsal ramus
206. ventral ramus
207. communicating rami
208. peruses
209. phrenic
210. brachial plexus
211. lumbosacral plexus

SOMATIC SYSTEM
212. true
213. true
214. true
215. true

AUTONOMIC SYSTEM
216. The autonomic system was originally thought to operate independently of the CNS.
217. Maintains homeostasis of peripheral functions which do not require conscious thought such as blood vessel diameter and heart rate.
218. Sensory signals pass directly to centers in the hypothalamus, brain stem, or spinal cord.
219. sympathetic and parasympathetic

220.	Sympathetic Autonomic System	Parasympathetic Autonomic System	Somatic Motor System (for comparison)
Number of neurons in pathway	2	2	1
Effector	visceral (smooth muscle, cardiac muscle, or gland)	visceral	skeletal muscle
Description of preganglionic neuron	originate only from 12 thoracic and first 2 lumbar nerves; myelinated; usually short compared to parasympathetic; most connect to sympathetic trunk through communicating rami	originate from brain stem and sacral nerves 2 to 4; myelinated; usually long compared to the sympathetic	not applicable
Description of postganglionic neuron	unmyelinated; extends from ganglion to visceral effector; long compared to parasympathetic	unmyelinated; extends from ganglion to visceral effector; short compared to sympathetic	not applicable
Location of ganglia	sympathetic trunk (along spinal cord) or prevertebral (in abdominal and pelvic cavities)	terminal ganglia (in or near effector)	not applicable
Neurotransmitter released by postganglionic fiber	norepinephrine	acetylcholine (ACh)	no postganglionic fiber, neurotransmitter is ACh

221. true
222. true
223. true
224. true
225. false, sympathetic
226. true
227. false, many
228. false, can be
229. false, hypothalamus and medulla oblongata

HOMEOSTASIS
Since you made up the question, you judge whether or not your answer is suitable.

CLINICAL TERMS
230. amyotrophic lateral sclerosis
231. Guillain-Barre syndrome
232. cephalalgia
233. bacterial meningitis
234. glioma
235. Guillain-Barre syndrome
236. encephalitis
237. schizophrenia
238. multiple sclerosis

LABELS AND LISTS
1. astrocytes, ependymal cells, microglia, oligodendrocytes, Schwann cells
2. multipolar, bipolar, unipolar. See your textbook for drawings of these three types.
3. dura mater, arachnoid mater, pia mater
4. meninges, vertebral column, cerebrospinal fluid (CSF)
5. forebrain, midbrain, hindbrain
6. flat bones of cranium, meninges, CSF
7. dura mater, arachnoid mater, pia mater
8. frontal, parietal, occipital, temporal
9. See your textbook.
10. See your textbook for the drawing. The resting potential of -70 mV is produced primarily by the action of the Na^+-K^+ pump. Depolarization occurs as Na^+ flows into the cell. Repolarization occurs as K^+ flows out of the cell. Resting potential is maintained through the action of the Na^+-K^+ pump.
11. - 16. See your textbook.

QUESTIONS TO MAKE YOU THINK
1. Na^+ is actively transported out of the cell while K^+ is actively transported into the cell by the cation of the Na^+-K^+ pump. During depolarization and repolarization ions flow by diffusion.
2. Muscle cells **contract** completely or not at all; whereas, nerve cells **conduct an impulse** in an all-or-none fashion.
3. Since three Na^+ are pumped out for every two K^+ into the cell, the net effect is a negative internal charge. Also K^+ diffuse out of the cell more readily then Na^+ diffuse into the cell, adding to the effect of the Na^+-K^+ pump.
4. The Na^+-K^+ pump as well as the ion channels which allow Na^+ and K^+ to flow into and out of the cell are integral membrane proteins. Since all of these proteins are involved in the transport of ions across the membrane, they must be part of the membrane, and likely cross the membrane themselves; thus, they are integral proteins.

ANSWERS - CHAPTER 10

SENSORY PATHWAYS
1. stimulus
2. receptor
3. threshold
4. stimulus-specific
5. sensory adaptation
6. mechanoreceptor
7. thermoreceptor
8. nociceptor
9. photoreceptor
10. chemoreceptor
11. general sensory pathway
12. three
13. special sensory pathway

GENERAL SENSES
14. cerebral cortex
15. cutaneous
16. Meissner's corpuscles
17. Pacinian corpuscles
18. visceral
19. referred
20. proprioceptor

SPECIAL SENSES - SMELL AND TASTE
21. a
22. b
23. d
24. b
25. b

VISION
26. d
27. d
28. a
29. d
30. b
31. a
32. b
33. c
34. d
35. c
36. d
37. d
38. b
39. d
40. c
41. d
42. b
43. a
44. d
45. d
46. c
47. c
48. a
49. b
50. c
51. b
52. d

HEARING
53. audition
54. auricle
55. auricle
56. tympanic membrane
57. external auditory canal
58. cerumen
59. tympanic cavity
60. auditory ossicles
61. Eustachian tube
62. Eustachian tube
63. mastoiditis
64. malleus
65. malleus
66. incus
67. stapes
68. oval window
69. labyrinth
70. labyrinth
71. vestibule
72. cochlea
73. perilymph
74. endolymph

75. bony labyrinth
76. perilymph
77. endolymph
78. scala vestibuli
79. scala tympani
80. organ of Corti
81. static equilibrium
82. dynamic equilibrium
83. otoliths
84. cristae

INTEGRATIVE FUNCTIONS
85. general sensory area
86. True
87. somesthetic association area
88. primary auditory area
89. True
90. gnostic area
91. consciousness
92. circadian rhythm
93. True
94. True
95. association
96. thought
97. short-term
98. False, there are two type of learning processes, short-term memory and long-term memory.
99. True
100. False, long-term memory results in anatomical changes within the brain.
101. limbic
102. False, prefrontal lobotomy is no longer used.

MOTOR FUNCTIONS
103. b
104. c
105. a
106. c, descending tract is white
107. b

CLINICAL TERMS
108. acataphasia
109. radial keratotomy
110. aphasia
111. Alzheimer's disease
112. achromatopsia
113. vertigo
114. eustachitis
115. dementia
116. cochlear implant
117. astereognosis
118. cataract
119. cerebrovascular accident
120. iridectomy
121. ametropia
122. glaucoma
123. epilepsy

MASTERY ESSAY
The tympanic membrane vibrates in response to sound waves passing down the external ear canal. These vibrations are transferred sequentially through the three auditory ossicles to the oval window. The movement of the stapes back and forth pushes the oval window in and out producing waves in the perilymph of the inner ear. These waves pass through the scala vestibuli and push the vestibular membrane inward, increasing the pressure of the endolymph within the cochlear duct. The waves in the perilymph are eventually dissipated through the round window into the tympanic cavity. The pressure waves within the endolymph cause the basilar membrane to move slightly. This movement bends the hair cells of the organ of Corti against the tectorial membrane. The bending of the hair cells releases neurotransmitters which, if present in sufficient quantity, initiate a nerve impulse in the cochlear nerve.

LABELS AND LISTS
1. mechanoreceptors - mechanical or physical change, thermoreceptors - temperature changes, nociceptors - pain, photoreceptors - light, chemoreceptors - chemicals
2. - 10. See your textbook.

Crossword Puzzle

Across:
1. SOMESTHETIC
3. VESTIBULE
5. MALLEUS
7. LABYRINTH
9. BITTER
11. MEMORY
13. STAPES
15. STIMULUS
18. OLFACTORY
19. SOUR
21. ASTIGMATISM
24. CONE
25. ADAPTATION
26. MECHANORECEPTOR
27. RECEPTOR
30. EYE
31. VISION
35. CORNEA
36. COCHLEA
37. IRIS
40. HYPEROPIA
41. COCHLEAR
43. SWEET
44. SENSATION
45. DEMENTIA

Down:
2. YPANIC (partial)
4. CEREBRAL
6. IMBIBIC
8. HEAT
10. INTEGRATION
12. OPSIN
14. SHOT
16. OLY
17. EPLEPSY
20. RHODOPSIN
22. ACCOMODATION
23. NOCICEPTOR
28. EHOFORROR
29. PUPIL
30. EYE
32. ENDOLYMPH
33. THOUGHT
34. GUSTATION
36. C
38. LINGUAL
39. INGO
42. ROD

ANSWERS - CHAPTER 11

COMPOSITION
1. homeostasis
2. hormones
3. slowly
4. glands
5. exocrine
6. endocrine
7. endocrine

HORMONES
8. hormones
9. target cell
10. water-soluble hormones
11. lipid-soluble hormones
12. lipid-soluble hormones
13. prostaglandins
14. water-soluble hormones
15. lipid-soluble hormones
16. cAMP
17. adenylate cyclase
18. protein kinases
19. protein kinases

HORMONAL CONTROL
20. In a negative feedback system, the effect of the hormone inhibits further secretion. For example, the parathyroid secretes PTH which causes Ca to be released from the bones into the blood. A rise in the Ca levels in the blood inhibits secretion of PTH.
21. In a positive feedback system, the effect of the hormone promotes further secretion. For example, oxytocin secreted by the pituitary causes the uterus to contract. Rising levels of oxytocin in the blood stimulate the secretion of more oxytocin. Positive feedback systems promote massive changes in physiology; therefore they are uncommon.
22. Some glands such as the adrenals and pituitary secrete hormones in direct response to nerve impulses. Secretion of norepinephrine is controlled in this manner.

PITUITARY GLAND
23. stimulates body cells to grow and divide
24. melanocyte-stimulating hormone (MSH)
25. prolactin (PRL)
26. adrenocorticotropic hormone (ACTH)
27. results in production and secretion of thyroid hormones
28. in females - stimulates development of ova and causes ovarian cells to secrete estrogens; in males - stimulates production of sperm
29. luteinizing hormone
30. oxytocin
31. regulates fluid balance
32. Regulating factors from the hypothalamus regulate the anterior lobe of the pituitary. The posterior lobe of the pituitary is an extension of the hypothalamus. It stores and releases hormones produced by the hypothalamus.
33. A tropic hormone is one which stimulates another endocrine gland. Gonadotropic hormones are a type of tropic hormone which stimulates gonads (sex organs).
34. growth hormone
35. hypoglycemia
36. hyperglycemia
37. gigantism
38. acromegaly
39. pituitary dwarfism
40. hypothalamus
41. oxytocin
42. antidiuretic hormone
43. osmoreceptors
44. antidiuretic hormone
45. antidiuretic hormone
46. growth hormone
47. melanocyte-stimulating hormone
48. prolactin
49. adrenocorticotropic hormone
50. thyroid-stimulating hormone
51. follicle-stimulating hormone
52. luteinizing hormone
53. oxytocin
54. antidiuretic hormone

THYROID AND PARATHYROID GLANDS
55. stimulate metabolism and promote growth
56. calcitonin (CT)
57. parathyroid glands
58. iodine
59. metabolism
60. protein
61. glucose
62. lipid
63. nervous
64. calcitonin
65. increases
66. calcitonin
67. uncontrollably excited

ADRENAL GLANDS
68. epinephrine and norepinephrine
69. adrenal cortex
70. maintain fluid balance
71. glucocorticoids
72. masculinizing and feminizing effects
73. d
74. a
75. d
76. b
77. c
78. b
79. d
80. c
81. d
82. d
83. d
84. a
85. b
86.

PANCREAS
87. pancreas
88. glucagon, insulin
89. islets of Langerhans
90. alpha cells
91. beta cells
92. glucagon
93. insulin
94. glucagon
95. insulin
96. Type I diabetes mellitus
97. Type II diabetes mellitus

GONADS, PINEAL GLAND, AND THYMUS GLAND
98. gonads
99. ovaries
100. testes (note: singular - testis)
101. estrogens, progesterone
102. testosterone
103. pineal gland
104. pineal gland
105. thymus gland
106. thymosin
107. thymus gland
108. thymus gland
109. thymus gland

HOMEOSTASIS

110. The nervous system senses changes in the environment and responds rapidly via nerve impulses, though these responses are short-lived and usually localized. Endocrine glands, on the other hand, respond by the secretion of hormones into the blood. Hormones can affect more cells, tissues, organs, etc. than individual nerve impulses, and these effects usually last longer, but are slower.
111. A balance output of hormones is necessary for homeostasis. Hyper secretion, over secretion, or hyposecretion, under secretion, can drastically affect the metabolic balance of the body. The exact effect, of course, depends upon the hormone and its target cells. For example, hypersecretion by the thyroid can cause increased metabolic rate resulting in weight loss and irritability.
112. There is no answer, as there is no question! You should be able to create Table 11-2 or any part of it from memory.

CLINICAL TERMS
113. diabetes insipidus
114. Cushing's syndrome
115. Addison's disease
116. amenorrhea
117. cretinism
118. myxedema
119. Grave's disease
120. exophthalmos
121. diabetes mellitus
122. goiter
123. diabetes mellitus
124. goiter
125. aldosteronism
126. Addison's disease

LABELS AND LISTS
1. pituitary, thyroid, parathyroids, adrenals, pancreas, gonads
2. pineal, thymus
3. kidneys, stomach, small intestine, placenta
4. change rates of: enzymatic activity, protein synthesis, secretion, transport across plasma membrane
5. pituitary, thyroid, parathyroids, adrenals, pancreas, gonads pineal, thymus
6. growth hormone, prolactin, thyroid-stimulating hormone, adrenocorticotropic hormone, melanocyte-stimulating hormone, follicle-stimulating hormone, luteinizing hormone, oxytocin, antidiuretic hormone
7. thyroxine, triiodothyronine, calcitonin
8. mineralcorticoids, glucocorticoids, sex hormones

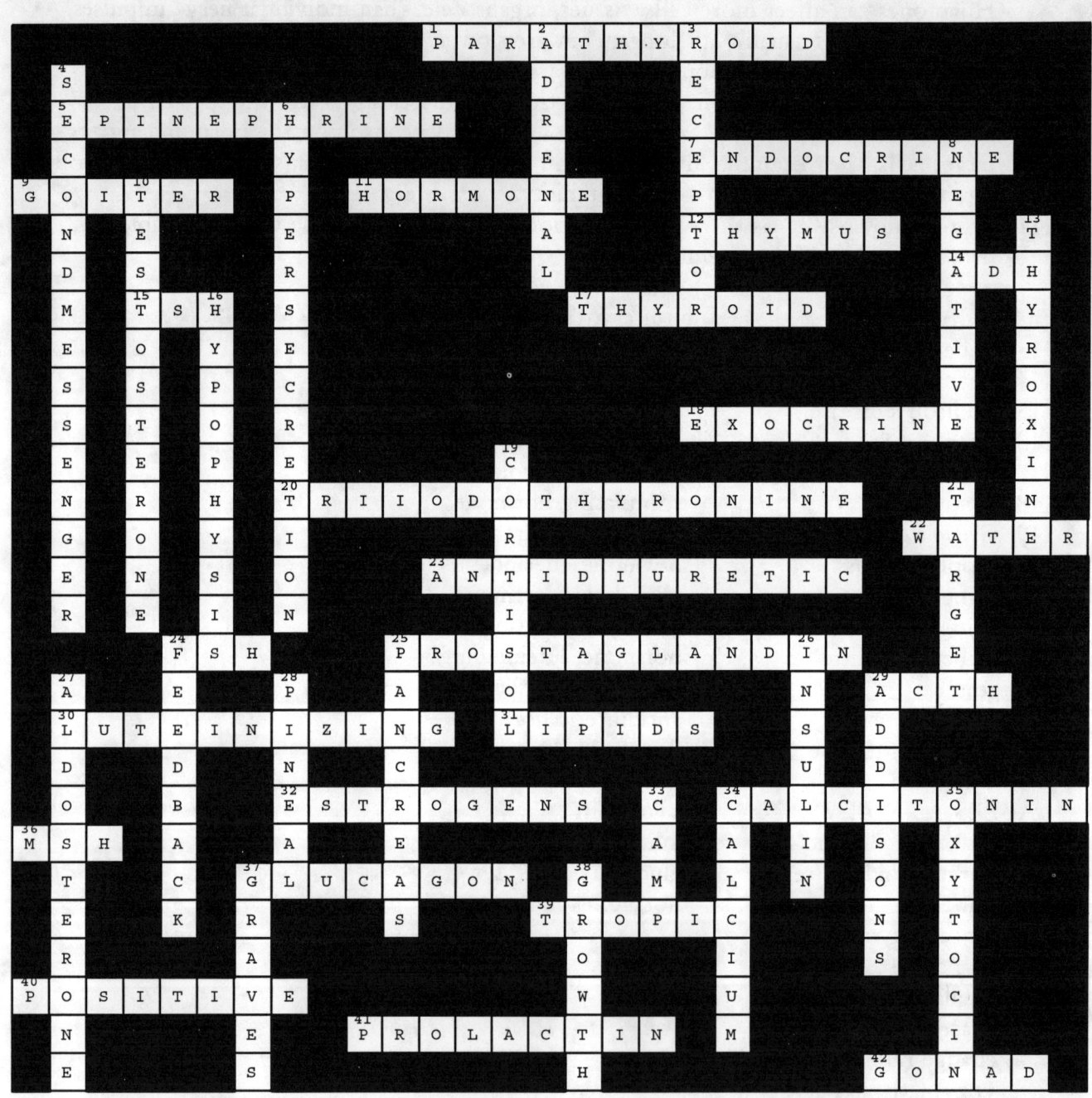

ANSWERS - CHAPTER 12

FUNCTIONS AND PROPERTIES OF BLOOD

1. circulatory system
2. cardiovascular
3. lymphatic
4. hematologic
5. formed elements
6. plasma
7. transportation
8. protection
9. regulation
10. antibodies
11. viscosity
12. pH
13. hemoglobin
14. true, except in pulmonary and fetal circulation
15. eight
16. acidosis (acidemia)
17. alkalosis (alkalemia)

PLASMA

18. true
19. true
20. false, water
21. true
22. false, increase
23. false, antibodies
24. false, clotting
25. true
26. true, though most oxygen is bound to hemoglobin
27. true
28. false, sodium

FORMED ELEMENTS

29. thrombocytes
30. erythrocytes
31. leukocytes
32. thrombocytes
33. hematocrit
34. blood smear
35. hematopoiesis
36. red bone marrow
37. hemocytoblasts, stem cells

ERYTHROCYTES

38. a
39. 40
40. c
41. b
42. c
43. d
44. b
45. d
46. a
47. d
48. c
49. c
50. d
51. a
52. a, b, c
53. b

LEUKOCYTES

54. white cell
55. control disease
56. yes
57. granulocytes, agranulocytes
58. neutrophil
59. eosinophil
60. basophil
61. monocytes, lymphocytes
62. neutrophil
63. eosinophil
64. basophil
65. lymphocyte
66. monocyte
67. between
68. diapedesis
69. they are attracted to chemicals released by microorganisms and damaged cells
70. leukocytes and plasma
71. antibodies
72. histamine
73. neutrophil & monocyte
74. facilitates movement of WBCs to the site of infection
75. antihistamines

PLATELETS

76. false, thrombocytes
77. false, have no
78. false, smaller
79. true
80. true

HEMOSTASIS

81. HEMOSTASIS
82. blood vessel spasm
83. serotonin
84. hemostasis
85. collagen
86. platelet plug
87. blood clot
88. fibrin
89. fibrinolysis
90. atherosclerosis
91. thrombus
92. embolus
93. E, D, A, C, B

BLOOD GROUPS
94. c
95. d
96. e
97. e
98. c
99. c
100. d
101. a
102. b
103. c
104. A
105. A, AB
106. B

BLOOD GROUPS CONT.
107. anti-A
108. B, AB
109. neither anti-A nor anti-B. It is improper to say that type AB has no antibodies, as this type would have antibodies against various microorganisms!
110. AB
111. A, B, AB, O
112. neither A nor B. It is inappropriate to state that type O has no antigens as the RBCs may have other antigens such as Rh.
113. O

HOMEOSTASIS
114. transportation, regulation, protection
115. The blood provides nutrients and oxygen to all body cells and carries away waste.
116. The immune functions of the blood protect the body from disease organisms.
117. The blood maintains the body's pH balance and assists in maintaining the temperature balance.

CLINICAL TERMS
118. hemorrhagic anemia
119. pernicious anemia
120. hemolytic anemia
121. hemorrhage
122. shock
123. sickle cell anemia
124. leukemia
125. hemophilia
126. bacteremia
127. anoxia
128. malaria
129. thalassemia
130. septicemia
131. leukocytosis
132. leukopenia
133. cyanosis
134. direct transfusion
135. apheresis
136. polycythemia
137. leukopenia

LABELS AND LISTS
1. erythrocytes, 4.2 - 6.2 million/mm^3; leukocytes, 5 - 10 thousand /mm^3; thrombocytes, 150 - 360 thousand/mm^3;
2. hemorrhagic, pernicious, hemolytic
3. granulocytes, agranulocytes
4. neutrophils, eosinophils, basophils
5. neutrophils, 55 - 60%; eosinophils, 1 - 4%; basophils, 0.5% or less; monocytes, 3 - 8%; lymphocytes, 35 - 33%
6. formation of prothrombin from thromboplastin and other substances, conversion of prothrombin in the plasma into thrombin, conversion of fibrinogen in the plasma into fibrin threads

QUESTION FOR THINKING
As a result of lack of protein in the diet, the child has too little plasma protein. There is therefore, a higher concentration of water in the plasma than is normal. Osmosis is the movement of water from higher concentration to lower concentration across a semipermeable membrane. Water moves from the blood into the abdominal cavity by osmosis. This causes the severe swelling that is seen.

439

ANSWERS - CHAPTER 13

CHARACTERISTICS AND COVERINGS
1. apex
2. pericardium
3. pericardium
4. pericardial sac
5. parietal pericardium
6. epicardium
7. epicardium
8. epicardium
9. pericardial cavity
10. pericarditis
11. fibrous layer of parietal pericardium

HEART WALL AND CHAMBERS
12. false, three
13. true
14. false, epicardium
15. false, myocardium
16. true
17. true
18. true
19. false, four
20. false, thinner
21. true
22. false, little ear
23. false, pectinate
24. true
25. true
26. true
27. true
28. false, little belly
29. true
30. false, right
31. true
32. true
33. papillary
34. true
35. false, coronary

HEART VALVES
36. tricuspid
37. three cusps
38. between right atrium and right ventricle
39. bicuspid or mitral
40. two cusps
41. between atrium and ventricle on the left side
42. pulmonary
43. three semilunar-shaped cusps
44. between right ventricle and pulmonary trunk artery
45. aortic
46. three semilunar-shaped cusps
47. The cusps of the bicuspid valve look like the miter hat of a bishop.
48. Blood pressure in the ventricles closes the valves. They are prevented from everting by the action of the papillary muscles pulling the valve via the chordae tendineae.
49. Blood can flow around the defective cusp creating a murmur.
50. Yes, the cusps of SL valves close due to blood pressure in the arteries. The margins of the cusps seal against one another to prevent back flow of blood.

BLOOD FLOW THROUGH THE HEART
See Figure 13-3 in your textbook. Remember that blood cannot be oxygenated (red) until it has gone to the lungs. Thus, blood in the pulmonary artery is "blue," whereas blood returning to the heart, via the pulmonary veins, is red. "Arteries are red, veins are blue, except in the lungs and some fetal too."

SUPPLY OF BLOOD AND HEART ATTACK
51. c
52. a
53. c
54. d
55. d
56. a
57. b
58. d

PHYSIOLOGY
59. cardiac cycle
60. systole
61. diastole
62. heart sounds
63. closing of the AV valves, ventricular systole NOTE: the sound is actually the closing of the valves, NOT the ventricular systole, though the two occur at the same time.
64. closing of the SL valves upon ventricular relaxation
65. sinoatrial (SA) node
66. atrioventricular (AV) node
67. bundle of His
68. Purkinje fibers
69. electrocardiogram
70. P wave
71. QRS complex
72. T wave
73. artificial pacemaker
74. heart rate
75. stroke volume
76. cardiac output
77. Starling's law of the heart
78. cardiac hypertrophy
79. acetylcholine
80. norepinephrine
81. baroreceptors

ARTERIES AND ARTERIOLES
82. away from NOTE: the color of an artery may be either red or blue. See the answer to number 50.
83. arteriole
84. tunica adventia
85. tunica intima
86. tunica adventia
87. tunica media
88. sympathetic
89. vasodilation
90. capillaries
91. one erythrocyte

CAPILLARIES
92. about 0.01 mm, about the size of a RBC
93. permit exchange of materials between the blood and body cells
94. thin walls
95. gas molecules, ions, and waste diffuse in or out of the capillaries
96. acts as a valve to control the flow of blood to capillaries
97. Capillary bed = branching network of capillaries; thoroughfare channel = capillary which connects directly to an arteriole or a venule; true capillary = capillary with capillary sphincter control, located between thoroughfare channels

VENULES AND VEINS
98. toward
99. venule
100. most
101. oxygenated
102. larger
103. distensibility
104. weakened valves
105. hemorrhoids
106. true
107. true
108. sinuses

BLOOD PRESSURE
109. c
110. b
111. a
112. d
113. d
114. b
115. c
116. d
117. c
118. b
119. d
120. b
121. d
122. d
123. d
124. a, d

CAPILLARY EXCHANGE
125. diffusion
126. osmotic pressure
127. hydrostatic pressure
128. venous end
129. arterial end

CIRCULATORY PATHWAYS
130. systemic circulation
131. pulmonary circulation
132. aorta
133. great saphenous vein
134. portal system

HOMEOSTASIS
135. condition in which body cells do not receive adequate blood, due to reduced cardiac output or blood volume; conditions: weak, rapid pulse, shallow and rapid breathing, pale, clammy skin, blue lips and fingertips, mental confusion and/or unconsciousness
136. renin, epinephrine, aldosterone, and ADH are secreted to raise blood pressure and volume

CLINICAL TERMS
137. heart fibrillation
138. aneurysm
139. congestive heart failure
140. murmur
141. tachycardia
142. ischemia or atherosclerosis
143. angina pectoris
144. pericarditis
145. stenosis
146. phlebitis
147. myocarditis
148. patent ductus arteriosus
149. hypertension
150. heart block
151. coarctation of the aorta
152. cardiac arrhythmias
153. bacterial endocarditis
154. atherosclerosis
155. arteriosclerosis
156. septal defects
157. aneurysm
158. bradycardia
159. tachycardia

LABELS AND LISTS
1. aorta, artery, arteriole, capillary, venule, vein, inferior vena cava
2. cardiac output, peripheral resistance, blood volume
3. right ventricle, pulmonary trunk, right (or left) pulmonary artery, pulmonary capillaries, pulmonary veins, left atrium
4. - 13. See your textbook.

QUESTIONS TO MAKE YOU THINK
1. Repolarization of the atria does involve electrical changes, but these changes are masked by the even greater electrical changes in the more massive ventricles. The atrial repolarization is hidden in the QRS complex.
2. Skeletal muscle tissue is striated, under voluntary nervous control, and functions primarily to move the skeleton. Cardiac muscle is striated, self-stimulating, and affected by the autonomic nervous system. Cells of cardiac muscle tissue are joined together by special junctions called intercalated disks. Smooth muscle is involuntary and non-striated. Smooth muscle forms the walls of blood vessels and visceral organs.
3. The P wave is caused by the electrical depolarization of the cells of the atria. This electrical event causes the muscle fibers to contract. The atria are systolic. The lub sound, produced by the closing of the atrial-ventricular valve, follows ventricular systole. Vomer is incorrect because the P wave is a graphical representation of an electrical event, which causes systole. The P wave, however, is not systole. The lub sound is a mechanical vibration, not an electrical event or contraction, and occurs at a different time from the P wave and atrial systole.
4. The muscle fibers of the AV bundle conduct an impulse from the AV node throughout the ventricular myocardium, much like a nerve impulse might travel down a neuron. However, the cells of the AV bundle are cardiac muscle cells, with striation, intercalated disks, etc. They are not neurons.
5. Cardiac muscle cells, including those of the SA node do not maintain a stable resting potential, but instead slowly depolarize after each repolarization. When the depolarization reaches a threshold level, the cells depolarize and produce action potentials which cause contraction.
6. Sodium and potassium are moved across the cell membrane by the sodium-potassium pumps to create a resting potential. As sodium gradually leaks into the cells, the potential is raised to threshold, sodium channels open, and an action potential is formed. The action potential spreads across the cell membrane and down the T-tubule membranes. The action potential causes the release of calcium ions from the sarcoplasmic reticulum. The calcium ions bind to troponin, resulting in structural changes that allow the myosin crossbridges to attach to the actin myofilaments resulting in contraction. The ions of sodium, potassium, and calcium are critical to the physiology of the heart.
7. Cardiac muscle cells are connected to one another physically **and electrically** by intercalated disks. When one cardiac muscle depolarizes, the depolarization spreads from cell to cell throughout the heart by means of the junctions of the intercalated disk.

ANSWERS - CHAPTER 14

LYMPHATIC NETWORK
1. d
2. b
3. a
4. c
5. b
6. a
7. d
8. a
9. a
10. b
11. d

OTHER LYMPHATIC ORGANS
12. lymphoid tissue
13. lymphocyte
14. afferent lymphatic vessels
15. hilus
16. efferent lymphatic vessels
17. fibrous capsule
18. lymph nodules
19. spleen
20. red pulp
21. white pulp
22. spleen
23. thymus
24. palatine tonsils
25. pharyngeal tonsils
26. lingual tonsils
27. Peyer's patches

DEFENSE MECHANISMS
28. self cells
29. nonself cells
30. MHC proteins
31. nonspecific defense mechanism
32. specific mechanism
33. skin
34. mucous
35. infection
36. phagocytosis
37. neutrophils
38. macrophage
39. macrophage
40. nonspecific defense mechanisms
41. natural killer
42. complement
43. interferons
44. inflammation
45. edema
46. true
47. Complement is a set of plasma proteins which acts to "tag" foreign particles and enhance phagocytosis.
48. Interferons bind to the membranes of cells and interfere with the ability of viruses to proliferate in the cells.
49. Redness is the result of vasodilation, part of inflammation.
50. Swelling is the result of vasodilation.
51. The added blood flow brings more leukocytes, antibodies, and other defensive substances to the site of infection or injury.

SPECIFIC MECHANISMS
52. lymphocytes, antigens
53. specific
54. cell-mediated
55. Antigens
56. autoimmune disease
57. immunoglobulins (Ig)
58. true
59. IgE
60. IgA
61. IgG
62. lymphocyte

63.

COMPONENT	DESCRIPTION	FUNCTION
Monocyte	large white blood cell, immature macrophage	identify, phagocytize, process, and present antigen
T cell	lymphocytes which mature in the thymus gland	active in cell-mediated immunity
Killer T cell	line of T cells produced after sensitization which	release lymphotoxins and destroy nonself cells
Helper T cell	line of T cells produced after sensitization which	release lymphokines which stimulate killer T cells, attract monocytes and neutrophils, and stimulate macrophages. Also stimulate B cells and humoral immunity.
Suppressor T cell	line of T cells produced after sensitization which	inhibit helper T cells
Memory T cell	line of T cells produced after sensitization which	store information on antigen structure and react rapidly to subsequent exposure
B cell	lymphocytes which mature in the red bone marrow or in Peyer's patches	active in humoral immunity
Plasma cell	line of B cells produced after sensitization which	produce antibodies
Memory B cell	line of B cells produced after sensitization which	store information on antigen structure and react rapidly to subsequent exposure
Antibody	protein molecules also called immunoglobulins produced by plasma cells	form specific chemical matches to specific antigens

64. antigen
65. macrophages (monocytes)
66. macrophages (monocytes)
67. B cells, T cells
68. killer T cells
69. killer T cells
70. helper T cells
71. memory T cells
72. suppressor T cells
73. B cells, plasma cells
74. macrophages
75. plasma cells
76. memory B cells
77. complement
78. antibody
79. natural killer cells

80. complement, antibody
81. B cells
82. macrophages
83. monocytes
84. B cells, T cells

85. c
86. d
87. a
88. b
89. a
90. a
91. b
92. c
93. a
94. d
95. a

ACQUIRED IMMUNITY

TYPE OF IMMUNITY	DESCRIPTION	EXAMPLES
naturally acquired active immunity	96. immune response stimulated by exposure to antigen	97. immune response to chicken pox
naturally acquired passive immunity	98. transfer of antibodies in a natural manner from one person to another	99. transfer of antibodies across placenta or through beast milk
artificially acquired active immunity	100. immune response stimulated by exposure to antigen which has been deliberately introduced into the body	101. vaccinations against measles, mumps, small pox, polio, etc.
artificially acquired passive immunity	102. human or animal antibodies injected into patient	103. treatment for rabies and snake bite

104. naturally acquired active immunity
105. naturally acquired passive immunity
106. vaccine
107. vaccination
108. naturally acquired active immunity
109. immune serum globulin
110. artificially acquired passive immunity

HOMEOSTASIS
111. false, can
112. true
113. false, unlike AIDS, cannot
114. false, older
115. true
116. false, is
117. false, is not (ARC is the precursor of AIDS)
118. false, eight to ten years
119. true
120. false, has not been
121. true
122. false, helper T cells
123. true
124. false, flu-like symptoms
125. true
126. false, and
127. false, feeble
128. true

CLINICAL TERMS
129. immunotherapy
130. lymphoma
131. neutropenia
132. acquired immunodeficiency diseases
133. anaphylaxis
134. immunization
135. DiGeorge's syndrome
136. Hodgkin's disease
137. monoclonal antibodies
138. vaccine
139. splenomegaly
140. autoimmunity
141. atopic diseases

LABELS AND LISTS
1. lymph nodes, spleen, thymus gland, tonsils, Peyer's patches
2. palatine tonsils, two; pharyngeal tonsils, two; lingual tonsils, two
3. vasodilation and increased permeability (seen as reddening and edema)
4. redness, swelling, heat, pain
5. IgG, IgA, IgM, IgE, IgD
6. killer, helper, memory, suppressor
7. plasma, memory
8. artificial active, artificial passive, natural active, natural passive,
9. homosexual males, intravenous drug users, hemophilia patients, sexual partners of above, fetus of infected mother
10. *Pneumocystis* pneumonia, Kaposi's sarcoma, hepatitis B, toxoplasmosis, weight loss due to chronic diarrhea (HIV wasting disease), AIDS dementia

449

450

ANSWERS - CHAPTER 15

1. b
2. b
3. c
4. d

ORGANS
5. upper respiratory tract, lower respiratory tract
6. conduction zone, respiratory zone

NOSE
7. nose
8. external nares
9. nasal septum
10. vestibule
11. nasal conchae
12. meati
13. sinusitis

PHARYNX
14. pharynx
15. skeletal muscles
16. mucous
17. internal nares
18. nasopharynx
19. oropharynx
20. laryngopharynx
21. oropharynx, laryngopharynx

LARYNX
22. larynx
23. larynx
24. larynx
25. larynx
26. thyroid
27. cricoid
28. epiglottis
29. thyroid cartilage
30. glottis
31. epiglottis
32. false vocal cords
33. true focal cords

TRACHEA
34. c
35. c
36. c
37. c
38. b

BRONCHIAL TREE
39. primary bronchi
40. bronchial tree
41. bronchioles
42. alveolar ducts
43. alveoli
44. surfactant
45. respiratory distress syndrome
46. respiratory membrane
47. asthma

LUNGS
48. right
49. apex
50. costal
51. medial
52. parietal pleura
53. visceral pleura
54. pleural cavity
55. right
56. segments
57. lobule

MECHANICS OF BREATHING
58. inspiration = breathing in, expiration = breathing out
59. into the area
60. primarily the contraction of the diaphragm assisted by external intercostal muscles
61. c, f, e, d, a, b
62. inspiration is the result of decreased air pressure in the lungs which causes atmospheric air to enter the lungs, much like a vacuum cleaner reduces air pressure and causes dust to move into the nozzle; expiration is the result of the elastic nature of lung tissue - the lung returns to its normal shape, reducing volume and increasing pressure which forces air out of the lungs
63. The pressure in the pleural cavity is slightly lower than that in the lungs so the lungs cannot collapse. Additionally, surfactant keeps the walls of the alveoli from sticking to themselves.
64. a, e, d, c, b
65. forced expiration follows the contraction of the internal intercostal muscles and is an active process, unlike the passiveness of normal expiration.

RESPIRATORY VOLUMES
66. spirometer
67. tidal volume (TV)
68. inspiratory reserve volume (IRV)
69. expiratory reserve volume (ERV)
70. residual volume (RV)
71. vital capacity(VC)
72. total lung capacity (TLC)
73. anatomic dead space volume

EXCHANGE OF GASES
74. c
75. a
76. c
77. a
78. b
79. b
80. c
81. a
82. a
83. c
84. b
85. b
86. b

CONTROL OF BREATHING
87. respiratory center
88. medullary rhythmicity center
89. pneumotaxic area in the pons
90. apneustic area in the pons
91. pneumotaxic
92. apneustic
93. chemosensitive area
94. chemoreceptors
95. hyperventilation
96. stretch receptors
97. SCUBA
98. rapture of the deep
99. bends

HOMEOSTASIS
100. The respiratory system must supply sufficient oxygen to the blood (to carry to the body cells) and expire waste CO_2. These functions must continue in times of exercise as well as in rest. To assure adequate gas exchange during exercise, breathing rate and depth increase. This is called hyperpnea.

CLINICAL TERMS
101. emphysema
102. dyspnea
103. asthma
104. apnea
105. anoxia
106. asphyxia
107. cystic fibrosis
108. hemothorax
109. pneumothorax
110. pulmonary embolism
111. pneumonia
112. pneumoconiosis
113. lung cancer
114. hypoxia
115. quinsy
116. tracheotomy
117. sudden infant death syndrome
118. rhinitis
119. sinusitis
120. respiratory infections

LABEL AND LIST
1. chemical changes in the blood, degree of stretch of the lungs, mental state
2. - 4. See your textbook.

ANSWERS - CHAPTER 16

ORGANIZATION
1. alimentary canal, accessory organs
2. gastrointestinal (GI)
3. 9 m (30 feet)
4. accessory

DIGESTIVE PROCESSES
5. ingestion
6. propulsion
7. peristalsis
8. swallowing
9. mechanical digestion
10. mastication
11. chemical digestion
12. absorption
13. defecation

SPECIAL FEATURES
14. peritoneum - serous membrane which supports many of the digestive organs
15. parietal peritoneum - covering of most digestive organs, consists of the falciform ligament, greater and lesser omenta, and mesentery
16. serosa (visceral peritoneum) - outer layer of the digestive organs
17. peritoneal cavity - potential space between the parietal and visceral peritoneum
18. falciform ligament - connects liver to abdominal wall and diaphragm
19. peritoneal folds - folds of the parietal peritoneum which support digestive organs
20. lesser omentum - connects stomach to liver and anterior abdominal wall
21. mesentery - connects the small intestine to the posterior abdominal wall
22. greater omentum - fatty apron descending from stomach over the small intestine, large number of lymph nodes provide immune protection
23. mucosa - inner layer (next to lumen) of alimentary canal
24. submucosa - layer external to mucosa (external = away from lumen), contains blood vessels
25. muscularis - muscular layer external to submucosa, causes peristalsis
26. serosa - outer covering of alimentary canal (also known as visceral peritoneum)

MOUTH
27. mastication
28. saliva
29. hard palate
30. fauces
31. uvula
32. vestibule
33. lingual frenulum
34. papillae
35. lingual frenulum

TEETH
36. twenty
37. thirty two
38. incisors
39. canines
40. molars
41. crown
42. enamel
43. periodontal ligament
44. dental caries
45. gingivitis
46. crown

SALIVARY GLANDS
47. a
48. d
49. a
50. a
51. b
52. c
53. a
54. c

PHARYNX ...
55. pharynx
56. internal nares
57. larynx
58. nasopharynx
59. oropharynx
60. laryngopharynx
61. bolus
62. tongue
63. soft palate
64. pharynx
65. epiglottis
66. peristalsis
67. esophagus
68. ten
69. esophageal hiatus
70. lower esophageal sphincter
71. heartburn

STOMACH
72. rugae
73. greater curvature
74. lesser curvature
75. pyloric sphincter
76. round bowl
77. gatekeeper
78. gastric pits
79. gastric glands
80. zymogenic or chief
81. parietal
82. mucous
83. mechanical
84. pepsinogen
85. HCl, intrinsic factor
86. mucus
87. pepsinogen
88. HCl
89. mucus
90. intrinsic factor
91. pepsin
92. gastric ulcer
93. chyme
94. hormones, involuntary control centers of the brain
95. gastrin
96. secretin, cholecystokinin

PANCREAS
97. acini
98. common bile duct
99. duodenum
100. endocrine (islets of Langerhans)
101. pancreatic amylase
102. pancreatic amylase
103. carboxypeptidase, chymotrypsin, trypsin
104. pancreatic lipase
105. pancreatic lipase
106. nucleases
107. secretin, cholecystokinin
108. secretin

LIVER
109. two
110. larger
111. liver lobules
112. hepatocytes
113. sinusoids
114. Kupffer cells
115. hepatocytes
116. bile salts
117. hepatic
118. bile
119. emulsification
120. glycogen
121. VLDLs
122. LDLs
123. HDLs
124. LDLs
125. cirrhosis of the liver
126. irreparable
127. protein metabolism
128. urea
129. hepatocytes
130. detoxified

GALLBLADDER
131. store and concentrate bile
132. hepatic duct drains the liver, cystic duct branches from the hepatic duct and feeds and drains the gallbladder, union of the cystic and hepatic ducts forms the common bile duct which fuses with the pancreatic duct (p. 905) and empties into the duodenum
133. sphincter of Oddi, relaxes with each wave of peristalsis

SMALL INTESTINE
134. small intestine
135. small intestine, 20
136. duodenum, jejunum, ileum
137. duodenum
138. jejunum
139. ileum
140. ileocecal valve
141. chemical and mechanical
142. intestinal villi
143. microvilli
144. surface area
145. lacteal
146. tuft of hair
147. intestinal glands
148. Peyer's patches (lymphatic nodes)
149. Brunner's glands
150. plicae circularis
151. true
152. lactose intolerance
153. 3, 10

LARGE INTESTINE
154. large intestine
155. 5
156. 3
157. cecum
158. vermiform appendix
159. colon
160. anus
161. appendicitis
162. peritonitis
163. anus
164. anal columns
165. taenia coli
166. haustrum
167. bile pigments
168. intestinal flora
169. intestinal flora
170. defecation
171. 18 to 24

172. fiber
173. If there are insufficient nutrients in the diet, plasma proteins are utilized by the body's cells. The loss of plasma proteins lowers the osmotic pressure of the blood and fluids accumulate in the interstitial spaces.
174. diverticulitis
175. achalasia
176. hepatitis
177. hemorrhoids
178. thrush
179. achlorhydria
180. aphagia
181. cholecystitis
182. colitis
183. hiatal hernia
184. gastritis
185. enteritis
186. pancreatitis
187. peptic ulcer
188. gastric ulcer
189. intestinal ulcer
190. cancers of the digestive organs
191. dysentery

LABELS AND LISTS

1. ingestion, propulsion, mechanical digestion, chemical digestion, absorption, defecation
2. falciform ligament, lesser omentum, mesentery, greater omentum
3. mucosa, submucosa, muscularis, serosa (visceral peritoneum)
4. tongue, teeth, salivary glands
5. incisors, canines, premolars, molars
6. parotid, submandibular, sublingual, buccal
7. laryngopharynx, nasopharynx, oropharynx
8. cardia, fundus, body, pylorus
9. mucosa, submucosa, muscularis, serosa
10. pepsinogen, HCl, mucus, intrinsic factor, pepsin
11. mechanical digestion, chemical digestion, absorption, propulsion, release of intrinsic factor
12. water, certain salts, glucose, alcohol, aspirin, lipid-soluble drugs
13. secretion of bile, carbohydrate metabolism (blood sugar homeostasis), fat metabolism, protein metabolism, storage of vitamins and iron, detoxification
14. long-term, heavy alcohol intake, prolonged exposure to heavy metals, heavy exposure to poisons
15. completes chemical digestion, primary site of nutrient absorption, propulsion
16. absorption of water, defecation
17. cecum, colon, rectum, anal canal
18. ascending colon, transverse coon, descending colon, sigmoid colon
19. unavailability of food, low nutrient value of food, defects in alimentary canal, anorexia nervosa, bulimia

ESSAY QUESTION
See pages 477 - 500.

20. - 26. See the figures in your textbook.

ANSWERS - CHAPTER 17

NUTRIENTS
1. food
2. produce energy or used as building blocks
3. process by which food is taken into the body and changed into a form which can be used
4. no, some nutrients are synthesized within the body from raw materials
5. a nutrient which cannot be synthesized

CARBOHYDRATES
6. false, sugar
7. true
8. false, from the liver and from ingested meat
9. false, monosaccharides
10. true
11. true

FATS
12. triglycerides
13. triglycerides
14. saturated fats
15. unsaturated fats
16. saturated fats
17. unsaturate fats
18. saturated fats
19. unsaturated fats
20. unsaturated fats
21. linoleic acid
22. cholesterol
23. cholesterol
24. 30%

PROTEINS
25. amino acids which cannot be synthesized by the body
26. 56
27. fat promotes cancer, fiber protects against cancer

VITAMINS
28. coenzymes
29. vitamins
30. fat-soluble vitamins
31. vitamin D
32. water-soluble vitamins
33. vitamin K
34. fat-soluble vitamins
35. water-soluble vitamins
36. water-soluble vitamins

MINERALS
37. bones, teeth
38. calcium, phosphorus
39. calcium, phosphorus
40. K, Na, Cl, Ca, P

BIOAVAILABILITY
41. the ability of the body to absorb only a portion of the vitamins and minerals ingested
42. combination of the nutrient with other nutrients, source of nutrients, body's needs
43. nutrients in foods can enhance bioavailability, whereas large doses in pill form may introduce imbalances or act as toxins
44. often do not supply necessary nutrients, may eliminate benefits and amplify harmful effects

TRANSPORT
45. monosaccharides
46. glucose
47. cellular respiration
48. liver
49. glycogen
50. bile salts
51. micelles
52. lacteals
53. chyle
54. lipoproteins
55. HDLs
56. LDLs
57. 180 mg/dl
58. stomach
59. active transport
60. intrinsic factor
61. protein
62. C
63. excreted
64. hypervitaminosis

METABOLISM
65. a
66. b
67. a
68. c
69. b
70. c
71. d
72. d
73. b
74. a
75. a
76. b
77. b

METABOLIC RATE AND BODY TEMPERATURE

78. the body' use of energy within a given period of time
79. the basic rate of chemical reactions within the body necessary to carry on life
80. the amount of energy required to carry out all body functions (BMR + voluntary muscular activity)
81. most of the energy released from food is in the form of heat
82. the level of humidity determines the ease with which H_2O evaporates, it evaporates much more easily in dry climates, thus cooling the body
83. these neurons respond to blood temperature, these impulses are relayed to blood vessels and sweat glands etc. which control body temperature

CLINICAL TERMS

84. heat stroke
85. hyperglycemia
86. phenylketonuria
87. obesity
88. hypoglycemia
89. fever
90. hyperglycemia
91. phenylketonuria

LABELS AND LISTS

1. carbohydrates, fats, proteins, vitamins, minerals, water
2. K, A, D, E
3. C, B_1 (thiamin), B_2 (riboflavin), B_6 (pyridoxine), niacin, pantothenic acid, biotin, B_{12}, folic acid
4. Ca, P, K, S, Na, Cl, Mg
5. combination of foods, source, need
6. nervous stimulation, hormones, body temperature, exercise, food intake
7. increased age, sleep, malnutrition, hypothermia
8. radiation = exchange of heat without physical contact, conduction = exchange of heat by direct contact, convection = transfer of heat between body and the air, evaporation = conversion of a liquid into a gas
9. vasoconstriction in the skin, skeletal muscle contraction, hormonal stimulation by thyroxine

ANSWERS - CHAPTER 18

KIDNEY STRUCTURE
1. retroperitoneal
2. liver
3. renal fascia
4. adipose capsule
5. renal arteries
6. 1,200ml/minute
7. 1,000,000
8. renal corpuscle
9. glomerulus
10. fenestra
11. podocyte
12. juxtaglomerular apparatus

KIDNEY FUNCTIONS
13. false, 45 (180 liters)
14. false, 1%
15. true
16. true
17. false, cannot
18. true
19. false, net filtration pressure
20. false, decreases
21. false, not
22. true
23. true
24. true
25. false, proximal
26. true
27. false, proximal convoluted tubule
28. true
29. true
30. false, nucleic acids
31. false, acute renal failure
32. true

REGULATION
33. renin
34. renin, angiotensin I, angiotensin II, aldosterone, increased blood volume, increased blood pressure, increase in filtration pressure
35. aldosterone
36. antidiuretic hormone
37. atrial natriuretic factor
38. urinalysis
39. bacteria
40. glucose

CONTROL OF URINE CONCENTRATION
41. juxtaglomerular cells of the kidneys
42. aldosterone
43. increased Na and Cl reabsorption, leading to H_2O reabsorption and more concentrated urine
44. ADH
45. atrial natriuretic factor
46. hypothalamus
47. vasoconstriction of afferent arterioles leading to reduced urine output

MAINTENANCE
48. they must be equal
49. kidney
50. 7.35 to 7.45
51. blood pH < 7.35, can result in nervous system collapse
52. blood pH > 7.45
53. chemical which resists a change in pH
54. increased breathing raises pH by faster elimination of CO_2
55. increase rate of secretion of H^+
56. reduce rate of secretion of H^+ and increase bicarbonate ion reabsorption
57. lowering of blood pH as a result of too little CO_2 being eliminated through lungs
58. excess production of hydrogen ions or loss of bicarbonate ions
59. excess loss of CO_2 due to rapid breathing
60. raising of blood pH due to loss of hydrogen ions

URETERS
61. b
62. c
63. c
64. b

URINARY BLADDER
65. urinary bladder
66. 1 pint
67. 2 + pints
68. 3
69. trigone
70. mucous membrane
71. detrusor

URETHRA
72. urethra
73. internal urethral
74. sphincter
75. external urethral
76. male

MICTURITION
77. micturition, urination, voiding
78. Stretching of the bladder wall stimulates stretch receptors which initiate impulses to the spinal cord which in turn send impulses to the detrusor muscle which pushes urine into the urethra. This is sensed as the urge to void.
79. contraction of the external urinary sphincter

HOMEOSTASIS
80. a, e, d, c
81. f, g, b, d, c

CLINICAL TERMS
82. cystitis
83. diuresis
84. glomerulonephritis
85. anuria
86. calculi
87. dysuria
88. enuresis
89. hematuria
90. urethritis
91. uremia
92. pyelonephritis
93. polyuria
94. oliguria
95. congenital polycystic kidney disease

LABELS AND LISTS
1. renal fascia, adipose capsule, renal capsule
2. afferent arteriole, glomerulus, efferent arteriole, peritubular capillaries
3. active transport, osmosis, diffusion, facilitated diffusion
4. renin, aldosterone, ADH, atrial natriuretic factor, sympathetic nerve impulses
5. buffers in blood, respiratory system, kidneys
6. - 9. See your textbook.

ANSWERS - CHAPTER 19

ORGANS OF MALE REPRODUCTION, TESTES
1. testes
2. testes
3. testosterone
4. sperm
5. scrotum
6. testes
7. seminiferous tubules
8. interstitial cells
9. interstitial cells
10. germ cells
11. spermatogenesis
12. primary spermatocytes, spermatogonia
13. spermatid
14. sterility

DUCTS
15. epididymis
16. six
17. peristaltic
18. flagellar motion
19. ductus deferens
20. inguinal canal
21. ejaculatory duct
22. spermatic
23. vasectomy
24. urethra
25. external urethral orifice

ACCESSORY GLANDS
26. b
27. a
28. c
29. a
30. d
31. a
32. b
33. c
34. c

EXTERNAL GENITALIA
35. scrotum
36. scrotum
37. dartos
38. penis
39. glans penis
40. corpora cavernosa
41. corpus spongiosum
42. corpus spongiosum
43. circumcision

PHYSIOLOGY
44. erection
45. emission
46. ejaculation
47. orgasm
48. anterior pituitary gland, hypothalamus, testes
49. gonadotropin-releasing hormone
50. gonadotropin-releasing hormone
51. luteinizing hormone, follicle-stimulating hormone
52. androgens
53. testosterone
54. puberty
55. testosterone
56. luteinizing hormone

ORGANS OF FEMALE REPRODUCTION
57. ovaries
58. ovarian follicles
59. oocyte, ovum, egg
60. menstrual cycle
61. primary oocytes
62. oogenesis
63. primordial follicle
64. twenty
65. fertilization
66. zygote
67. antrum
68. estrogen
69. Graafian follicle
70. ovulation
71. corona radiata
72. corpus luteum
73. corona radiata
74. uterine tube (fallopian tube)
75. corpus luteum
76. corpus luteum

FEMALE ACCESSORY ORGANS
77. fallopian tube, oviduct
78. infundibulum
79. infundibulum
80. fimbriae
81. uterus
82. cervix
83. vagina
84. serous, myometrium, endometrium
85. endometrium
86. vagina
87. vestibular glands
88. vaginal orifice

EXTERNAL GENITALIA
89. vulva
90. mons pubis
91. labia majora
92. clitoris
93. vestibular glands
94. perineum
95. episiotomy
96. labia majora
97. vestibule

MAMMARY GLANDS
98. Increasing levels of estrogen and progesterone cause the mammary glands to enlarge.
99. Alveolar glands are the milk-secreting glands of mammary tissue. They are contained in small chambers called lobules. Lobules are arranged into lobes which make up the mammary glands or breasts.

PHYSIOLOGY OF
100. hormones
101. ovulation
102. menstruation
103. pregnancy
104. sexual stimulation
105. clitoris
106. mucus
107. vestibular glands
108. sperm
109. hypothalamus
110. anterior pituitary
111. ovaries
112. estrogen
113. gonadotropin-releasing hormone (GnRH)
114. hypothalamus
115. anterior pituitary gland
116. follicle-stimulating hormone (FSH), luteinizing hormone LH)
117. ovary
118. estrogen, progesterone
119. estrogen, progesterone
120. menarche (first menstruation)
121. changes
122. fat
123. progesterone
124. corpus luteum
125. menstrual
126. menses
127. endometrium
128. estrogen
129. proliferative
130. GnRH
131. FSH, LH
132. ovulation
133. corpus luteum
134. progesterone
135. rise
136. endometrium
137. secretory
138. endometrium
139. corpus luteum
140. dysmenorrhea
141. prostaglandins
142. premenstrual syndrome
143. climacteric
144. menopause
145. contraception
146. vasectomy

CLINICAL TERMS
147. gonorrhea
148. herpes
149. hysterectomy
150. pelvic inflammatory disease
151. toxic shock syndrome
152. vaginitis
153. amenorrhea
154. breast cancer
155. endometriosis
156. cancer of the prostate
157. cervical cancer
158. cryptorchidism
159. dilation and curettage
160. fibrocystic disease
161. fibroadenoma
162. syphilis
163. trichomoniasis
164. vaginitis
165. hysterectomy
166. dilation and curettage
167. toxic shock syndrome
168. syphilis

LABELS AND LISTS
1. epididymis, ductus deferens, ejaculatory duct, urethra
2. ductus deferens, testicular artery, testicular veins, lymph vessels, testicular nerve, cremaster muscle, connective tissue
3. seminal vesicles (2), prostate gland (1), bulbourethral glands (2)
4. hypothalamus, anterior pituitary, testes
5. GnRH, LH, FSH, estrogen, progesterone
6. uterine tubes, uterus, vagina
7. - 10. See your textbook.

465

ANSWERS - CHAPTER 20

PRENATAL DEVELOPMENT
1. embryology
2. postnatal development
3. fertilization
4. embryonic period
5. fetal period
6. fertilization
7. 1 hour
8. 24 hours
9. hyaluronidase
10. corona radiata
11. haploid
12. haploid
13. diploid
14. zygote
15. infertility
16. *in vitro*

EMBRYOLOGY
17. mitosis
18. cleavage
19. uterine tube
20. morula
21. trophoblast
22. embryo
23. blastocoele
24. blastocyst
25. implantation
26. endoderm
27. mesoderm
28. ectoderm
29. chorion
30. amnion
31. gastrulation
32. gastrula
33. neurulation
34. foregut
35. umbilical cord
36. amniotic
37. organogenesis
38. 21
39. seventh
40. chorionic villi
41. two arteries, one vein
42. months 4-9
43. is
44. can
45. fetus
46. lanugo
47. 38

PARTURITION
48. Parturition is the birth process.
49. Braxton-Hicks contractions are small contractions of the uterine wall due to increasing levels of estrogen from the placenta. These minor contractions prepare the uterus for the more powerful birth contractions to follow.
50. Stretching of the uterine wall as the fetus grows signals the hypothalamus to stimulate the posterior pituitary to secrete oxytocin which causes the myometrium to contract.
51. Labor refers to the increased, rhythmic contractions induced by oxytocin which lead to birth.
52. The contractions cause the amnion to break, releasing the amniotic fluid surrounding the fetus. This is known as "breaking water" or "water bag rupture".
53. Cervical dilation refers to the stretching of the cervix due to the contractions.
54. The first stage of labor is cervical dilation to about 10 cm.
55. The second stage of labor involves the actual expulsion of the fetus.
56. The third stage of labor involves the expulsion of the placenta and umbilical cord.

POSTNATAL DEVELOPMENT
57. b
58. b
59. c
60. b
61. a
62. d
63. a, c, d
64. d
65. d

GENETIC INHERITANCE
66. genetics
67. 46
68. homologous
69. alleles
70. homozygous
71. heterozygous
72. genotype
73. phenotype
74. Aa
75. aa (or AA if it were a choice)

DOMINANT-RECESSIVE INHERITANCE
76. A dominant gene (or allele) is one which is expressed as a phenotype. A recessive allele is one which is hidden by a dominant allele and is in the phenotypic expression only if it is homozygous (aa for example).
77. A Punnett square is a chart containing four or more boxes used to assist in solving genetic problems.
78.

	f	f
F	Ff	Ff
f	ff	ff

Ff = freckled
ff = non-freckled

SEX-LINKED INHERITANCE
79. sex chromosomes (X, Y)
80. autosomes
81. X
82. X and Y
83. male
84. female
85. sex-linked
86. X-linked
87. Y-linked
88. no

GENETIC SCREENING
89. A small mount of amniotic fluid is withdrawn through a needle and tested for genetic abnormalities.
90. tests for phenylketonuria, several types of retardation, and hemophilia

CLINICAL TERMS
91. miscarriage
92. senility
93. abortion
94. Klinefelter's syndrome
95. fetal jaundice
96. ectopic pregnancy
97. cesarean section
98. amniocentesis
99. hydatid mole
100. meconium
101. Turner's syndrome
102. placenta previa
103. preeclampsia
104. tubal pregnancy
105. congenital defects
106. hydramnios
107. septal cardiac defects

LABELS AND LISTS
1. zygote, morula, blastocyst, gastrula, neurula, embryo (all of the above are embryonic stages as well), fetus, neonate
2. capillary wall, basement membrane, chorion
3. foramen ovale becomes the fossa ovalis, ductus arteriosus, umbilical vein becomes part of the falciform ligament, ductus venosus becomes a solid ligament, umbilical arteries